D1333371
COLLEGE LIBRARY
00016541

creatures of habit

understanding African animal behaviour

creatures of habit

understanding African animal behaviour

text by Peter Apps

photographs by Richard du Toit

ACKNOWLEDGEMENTS

PETER APPS

Writing the words was only part of the process that put this book into your hands, and I would like to thank Louise Grantham, Renee Ferreira, Frances Perryer, Pippa Parker, Janice Evans and Helena Reid for their inputs at various stages of the book's conception, gestation, birth and development.

Richard du Toit went out of his way to get the photographs that were needed to make a unity of text and illustrations. The results of his effort speak for themselves.

And thank you, Helen, for immersing yourself in your books while I was immersed in this one.

RICHARD DU TOIT

Many people helped me in various ways to obtain the photographs for this book. To everyone not mentioned below, I offer my sincere thanks.

Firstly, a special word of thanks to my good friend Gerald Hinde for his great help, inspiring ideas, and encouragement over the years.

To Will Taylor, Linda Slaughter and Craig Kissick of Panthera Productions in Dallas, Tony and Sharon Heald, Molly Buchanan, Concetta Serretta, Wayne and Vanessa Hinde, and Royston Knowles – a big thanks to all of you for your assistance.

It was a pleasure working with the talented and enthusiastic people at Struik, especially Pippa Parker and Janice Evans.

To Bobby Haas of Dallas, thanks for those great photographic adventures and for your special contribution to cheetah and wildlife conservation in Africa.

And finally to Mom, Dad, Judith, Elizabeth, Justin and Alan – thanks for everything.

PHOTOGRAPHIC NOTES

The photographs in this book have not been manipulated or altered in any way by computer software. These images are an accurate record of the behaviour of animals in their natural habitats.

All the photographs in this book were taken with ISO 50 and 100 slide film using a 35 mm SLR camera system. I used camera bodies with power boosters and several different lenses. I used two zoom lenses, the 17–35 mm f2.8 and the 70–200 mm f2.8, while my long lens is a 500 mm f4.5. I occasionally used a flash for daytime fillflash and low light conditions, and also for normal night photography.

I buy all my film and accessories from Kameraz in Rosebank, Johannesburg. They are the biggest sellers of second-hand camera equipment in South Africa, and are currently expanding into digital photography, both hardware and software ranges.

RICHARD DU TOIT

Struik Publishers
(a division of New Holland Publishing (South Africa) (Pty) Ltd)
Cornelis Struik House
80 McKenzie Street
Cape Town 8001
www.struik.co.za

New Holland Publishing is a member of the Johnnic Publishing Group.
Log on to our photographic website
www.imagesofafrica.co.za for an African experience.

First published in 2000
10 9 8 7 6 5 4 3 2

Publishing manager: Pippa Parker
Managing editor: Simon Pooley
Designer: Janice Evans
Editor: Frances Perryer
Co-ordinator: Helena Reid
Proofreader: Annelene van der Merwe
Indexer: Claudia dos Santos

ISBN 1 86872 433 6

Reproduction by Hirt & Carter Cape (Pty) Ltd
Printed and bound by Craft Print International Ltd, Singapore

Front cover: Lioness with cub
Back cover: (clockwise from top left) Elephant in threat pose, blue wildebeest with calf, springbok pronking, zebras fighting.
Half title: Behaviour links animals to their environment; a steenbok's selection of only the most nutritious parts of its food plants allows it to penetrate a variety of habitats.
Full title: Classic Africa. Understanding animal behaviour adds an extra dimension to the wildlife experience.
Endpapers: Survival in action – fleeing impala confound a predator with an explosion of agile grace.

CONTENTS

CHAPTER ONE

INTRODUCTION

mammals and behaviour

EVOLUTION

ETHOLOGY

SENSES

LEARNING

Previous pages: Mammals are uniquely flexible in their behaviour – elephants will drink and bathe at least once a day if they can, but if they have to, they can go for three or four days without water.

Above: Animals never do nothing; by submerging in water hippos protect themselves from both the sun and predators.

MAMMALS OCCUPY A BROADER RANGE OF HABITATS and pursue a wider range of lifestyles than any other group of animals. They graze, browse, hunt, scavenge and filter-feed, and consume fruit, invertebrates, grass, leaves, gum, bark, fish, reptiles, crustaceans, roots, seeds and each other. They live on land, up trees and down burrows; they swim in fresh water and in the sea, and they fly through the air. There is no animal group with a wider range of body sizes – from 1,5 gram Kitti's hognose bats to 120 tonne blue whales, the largest animals ever to have lived on Earth. The largest modern land animal (the African elephant) is a mammal; the hunters' big five (elephant, rhino, leopard, lion and buffalo) and African ecotourism's big seven (elephant, rhino, leopard, lion, buffalo, cheetah and wild dog) are mammals, and we are mammals ourselves – we are warm-blooded creatures with hair, and we were nourished with milk by our mothers.

Behaviour is the things that animals do – from the violent struggle of a lion pride pulling down a buffalo bull, to the delicate precision of a mouse collecting seeds; from the stern glance with which a dominant male baboon displaces a subordinate from a shady resting spot, to the horn-clashing battle of kudu bulls fighting for social rank and the access to females that it confers. Some behaviours are universal; others appear only in animals of certain species, only in one of the sexes, or only in animals of a certain age. Yet others appear when rare and special sets of circumstances call them up. Within each individual, hunger, thirst, fear, aggression and a host of other motivations interact with one another and with experience, expectations and stimuli from the environment to trigger, inhibit or modify animal behaviour.

Animals never do nothing: a mouse sleeping in its burrow, a hippo dozing in the shallows of a river, and a gemsbok standing motionless in the shade of a Kalahari camel-thorn tree are all doing the same thing – using behaviour to help regulate their body temperatures, which an elephant does more actively by fanning its ears and spraying itself with water, and a ground squirrel by lying spreadeagled on the ground and showering cool sand onto its back. The same mouse in its burrow is also as surely avoiding predators as an antelope that stops feeding every now and then to lift its head and scan for approaching danger.

Because mammals and their behaviour are so diverse and so complex, there is no such thing as typical mammal behaviour – the typical mammal, like the average man, is a mythical beast. At a finer level, there is no single individual whose behaviour can stand for a whole species. To talk about *the* lion or *the* elephant is a relic of the days when it was thought that all animals of a species varied within such narrow limits that to study one individual was to obtain results that were relevant to all. That presumption is less valid for behaviour than for any other aspect of biology, because behaviour varies more than any other feature, both from one animal to another and within individuals. And not only do individuals behave differently, but the differences are critical components of social structures – only within the most amorphous of animal aggregations, such as shoals of fish or large flocks of birds, does each member have an anonymous, interchangeable role. In the complex social groups that are so characteristic of mammals, individual roles that have been shaped by age, sex, experience, status and individual history are the very building blocks of social structure.

By generating body heat internally, mammals maintain a high and fairly constant body temperature which allows their behaviour to be relatively independent of the temperature of their surroundings. The rapid metabolism needed for this endothermy demands efficiencies in feeding,

A lioness's closely meshing set of teeth of different shapes and sizes is a uniquely mammalian feature.

digestion and respiration which mammals achieve through their finely tuned senses, speed and agility, sophisticated teeth and a hard palate that separates the breathing passages from the mouth. A coat of hair provides insulation, and body temperature is controlled by feedbacks to metabolic rate, behaviour and cooling mechanisms such as sweating.

Behaviour is the interface between an animal and its environment, and understanding the behaviour of individuals is essential to understanding the ecology of populations, habitats and ecosystems. A carnivore that captures its own food can influence the population dynamics of its prey, but if the same carnivore feeds on carcasses of dead animals it can have no such effect. A predator that removes adults from prey populations has more effect than one that takes juveniles. In the Kruger National Park, for example, lions have a greater impact on blue wildebeest numbers than they do on Burchell's zebras because they prey mainly on zebra foals and on wildebeest adults.

Mammal societies, like this baboon troop, are complex webs of relationships between different individuals.

Elephants' feeding behaviour makes them a major force in the shaping of habitats and ecosystems.

What separates a grazer from a browser is the behaviour of food selection – and mammals of many species can switch from one behaviour to the other depending on season or food availability. Elephants are only one of several large mammals that eat the seeds, leaves and young branches of trees; what sets them apart and makes them such powerful agents for habitat modification is their feeding behaviour – where a giraffe, kudu or black rhino feeds by cropping the edible parts, leaving the tree more or less intact, an elephant demolishes a whole tree, and a herd of elephants a whole woodland. Management of elephant populations and their habitats demands an understanding of their behaviour, not just of their numbers.

EVOLUTION

Animals have features that seem to have been designed for a purpose: bats have wings for flying, hyaenas have powerful teeth and jaws for crushing bones, and male antelope have horns for fighting. Features like these, which enable organisms to face and overcome the challenges of survival, reproduction and raising their young, and which appear to have been designed to fulfil that function, are known as adaptations. The best scientific explanation for the adaptation of living organisms is that it is a result of natural selection – those individuals that had features better suited to their environments produced more offspring than others, and their offspring inherited the same sound design features and passed them on, through genetic inheritance, to their offspring in turn. In each generation, those individuals that carried genes for the better-suited features passed on more genes by successfully raising more offspring, and so those genes increased in frequency at the expense of those that led to poorer designs. The result is that animals now carry genes for doing things well – for being well adapted.

The basis of evolutionary ethology is that, all else being equal, we would expect to see animals behaving in ways that increase the number of offspring that they can rear to independence with a good chance of having offspring of their own – in other words, doing things that contribute to their success at bequeathing copies of their genes to subsequent generations.

There is a broader view of the evolution of behaviour that brings apparently altruistic behaviour such as the sentinel systems of dwarf mongooses and suricates (Chapter 5) into line with the rest of evolutionary biology. Close relatives share a genetic heritage – therefore an animal that aids a relative's reproductive efforts is at the same time helping to bequeath some of its own genes to subsequent generations. At the genetic level, it is reproducing by proxy. Close kin share more genes than do more distant relatives, and the phrase 'kin selection' has been coined for the process whereby helping close relatives to reproduce has evolved in the face of natural selection at the level of individual animals.

Because the natural selection that drives evolution operates at the level of individual animals or their closest relatives, it would be very surprising to find an animal doing anything 'for the good of the species'. One that did so instead of looking to its own genetic interests would fare less well against natural selection than one that took the 'selfish' course. And, indeed, we do not see animals behaving for the good of their species; even less do we see them doing anything that curtails their reproduction but benefits their habitats or the ecosystems that contain them. As only one example, elephants show no restraint in their breeding except what is imposed on them by their physiological reactions to starvation and stress, which are themselves the results of the environmental devastation wrought by elephant populations that have reached high density.

Natural design – a hyaena's bone-crushing jaws allow it to exploit carcasses that other carnivores cannot.

Understanding the evolution of animal behaviour requires more subtle modelling than simple population dynamics and population genetics. While physical features can vary only within very narrow limits, behaviour is much more flexible – a leopard cannot change its spots, for example, but it can very easily change its habits. And culture can transmit behaviour from generation to generation without the need for genetic inheritance.

To understand why white rhino bulls are prepared to fight to the death over territory, we have to know that only territory holders get access to females.

ETHOLOGY

The study of animal behaviour stretches back in time past the domestication of our farm and companion animals into the most distant beginnings of the human race, when our hunter-gatherer ancestors needed an intimate familiarity with the habits of their prey and predators in order to eat and avoid being eaten. Despite these ancient beginnings, the study of animal behaviour as a truly scientific discipline dates back less than 100 years.

Although 'ethology' as a label has recently been hijacked by pet psychologists looking for an up-market image, the word was first used in 1854 to label the study of how an animal's structure is related to its habits. Ethology as a scientific discipline has now come to emphasise habits in relation to habitats and structures. As well as asking what an animal does, when, where, how, and with whom, ethology asks and tries to answer another question – 'Why?'. That seemingly straightforward enquiry is multidimensional, with links to laboratory studies of animal psychology on the one side and to ecology, palaeontology, population genetics and evolution on the other.

To be properly considered as science, the answers to the question of why an animal behaves as it does have to be logical, and have to relate to what goes on in the real world. If the predictions and conclusions that follow from our explanations do not match what really happens, we can be sure that no matter how elegant, persuasive, all-encompassing and fervently advocated our ideas are, they are somehow wrong. Like all sciences, ethology progresses by discarding faulty explanations and replacing them with alternatives that fit the world more closely.

Why does an animal behave in a particular way at a certain time? That question can be asked, and answered, in four ways. First, the question of causation – what stimulus from the environment or change within the animal itself triggers a particular behaviour? Second, the question of development – is the behaviour innate or did the animal learn how to do it; is it hard-wired into the nervous system or programmed in by learning? Third, the question of function – what is the behaviour for, what purpose does it serve, how does it contribute to the animal's survival and reproductive success? And fourth, the question of evolution – what ancestral behaviours were at the roots of what we see today, and along what evolutionary branches did they grow into their present form?

Of course, the 'Four Whys' are intimately interwoven. A hungry leopard is more motivated to hunt than one that is well fed. If it sights a potential victim it will begin a stalk, whose function is to bring the leopard within striking range of its prey. Stalking is a typical hunting tactic for cats: it has evolved in concert with their lithe and muscular builds, their inability to sustain high-speed chases, their solitary habits, and their preference for habitats that provide good cover. The progress of any particular hunt will be an amalgam of those innate, evolved strategies with tactical responses to the terrain and the prey's behaviour, which the leopard will have learned from experience.

In contrast to a solitary leopard, within its highly social pack an individual wild dog's motivation to hunt depends on its packmates' behaviour as well as on its own hunger. The pack's hunting methods – long, high-speed chases that match the stamina of the hunter against that of the hunted – have evolved from those of a canine ancestor that also ran down large prey in open country.

An animal's behaviour at a given moment is determined by more than the immediate set of circumstances; it is also shaped by the experiences that the animal has already lived through, and by its expectations of what is to come (although this by no means implies that animals can see into the future). Its expectations have been moulded by natural selection: over hundreds or thousands of generations its ancestors have been the ones whose behaviour came the closest to the ideal of preparation for future events. A spotted hyaena male working his way into a new clan behaves as if he knows that cringing submissiveness in the face of aggression from the clan females will provide him with future opportunities to mate, but only because that was the strategy that earned his father, grandfathers and great-grandfathers those opportunities, in a line stretching back for hundreds of thousands of generations.

A leopard's hunting behaviour depends on complex interactions between motivation, development, function and evolution.

In some cases, the function of a behaviour is obvious. Bachelor herds are obviously not mating aggregations, but they do offer the protection against predators afforded by extra pairs of eyes and ears (*see* Chapter 4). Other clues to the function that a behaviour serves can be gleaned from comparisons between species. For example, we might be interested in why large predators such as lions and spotted hyaenas live in groups – what is the function of such sociality; what problem of survival or reproduction does it help social carnivores to solve? If we compare the diet and hunting success of lions with those of leopards, which are typical feline solitaries, we find that lions are more successful than leopards at preying on large herbivores: lions score a 15–30 per cent success rate, while leopards are successful in only five per cent of hunts. There is an even more convincing difference between solitary brown hyaenas and social spotted ones: the browns get a maximum of five per cent of their food from hunting and never catch anything bigger than a springbok lamb, while a clan of spotted hyaenas will hunt for at least 70 per cent of its food, and is perfectly well able to bring down adult zebras and gemsbok. These differences suggest that the ability to tackle large, dangerous prey might be why lions and spotted hyaenas live in groups. On the other hand, it could just as well be that their ability to capture large prey is due only to their being large themselves – and there are no solitary hunters of the same size with which we can compare them.

More persuasive evidence than is provided by comparisons between species comes from comparing the performance of solitary and grouped animals of the same species. More persuasive still are the variations in success that a particular individual experiences when it joins a group or operates alone. For example, in the Ngorogoro crater spotted hyaenas hunt for small, defenceless Thompson's gazelle fawns in pairs, which is five times as effective as hunting alone, but form groups of up to 27 when in pursuit of zebra, which are larger, faster and much more dangerous. Among lions, the proportion of hunts that are successful rises with the number in the hunting group, but it does not rise fast enough to compensate for the larger number of mouths that have to be fed from the spoils. In a large group each member actually gets less meat, both per meal and averaged over time, than it would in a smaller group. But the enhanced hunting prowess of a large group means that meals come more regularly and more dependably.

The more lions there are in a pride, the more thinly the spoils of the hunt are spread among the members.

The fourth why – how did behaviour patterns evolve into their present form? – is probably the most difficult to answer, because there is almost no evidence to work with. Apart from the odd fossilised footprint, some tooth marks on fossil bones, and what can be deduced from the sizes and shapes of fossilised skeletons, behaviour has left no traces in the fossil record from which to reconstruct the habits of modern animals' extinct ancestors. Although fossil skulls show that bats have been using echolocation (*see* Chapter 3) for at least 50 million years, we do not know when lions or their ancestors started living in prides, or when an ancestral suricate first climbed onto a termite heap to watch for approaching predators.

Because behaviour has such a strong time dimension, the most fruitful studies of behaviour are those that go on long enough to follow animals all the way from birth, through infancy and adulthood, to death. Extend such studies across several generations and the insights that they generate become more and more enlightening, the questions that they pose more and more intriguing. In a baboon troop, a female's social rank is passed on to her daughters, but not to her sons, who disperse

Above: Hunting in pairs or larger groups increases spotted hyaenas' success rate.

Opposite, top and middle: Both predators and prey need acute night vision. Eye shine is light that has been reflected back through the retina of the eye to increase visual sensitivity.

Opposite, bottom: A bushbaby's huge eyes make starlight bright enough to see by.

to new troops where their rank depends on their individual abilities rather than on the blood-line alliances among the females that remain in their natal troops. Observations of male lions killing cubs made sense only when long-term studies revealed that they killed only cubs that had been sired by other males, and that their own reproductive success increased as a result (*see* Chapter 5).

Of the three founders of ethology – Karl von Frisch, Konrad Lorenz and Niko Tinbergen – it is probably Tinbergen whose example has most significantly influenced the development of the science into its present form. In addition to formulating the 'Four Whys', he nurtured the conviction that a real understanding of an animal's behaviour has to be based on studies of how it lives in its natural habitat. Paradoxically, when Tinbergen's advice was followed, it was the field studies he advocated that prompted a move beyond the boundaries of classical ethology by showing that behaviour varied almost as much within species as between them, challenging one of the tenets of classical ethology – that species-specific behaviour was as invariant as a coat pattern or the shape of a horn.

Research into animal behaviour draws on such a diversity of investigative techniques that the traditional divisions between scientific disciplines have blurred and broken down. Specialists have had to become generalists, and then become experts in new fields. For example, the DNA fingerprints provided by molecular genetics draw pictures of family bloodlines far more accurate than the most expert and dedicated of field workers' observations of who mates with whom, and which are the offspring of what unions. To study the chemistry of mammals' scent signals pushes to the limits the expertise of analytical chemists and the sensitivity and accuracy of their instrumentation. Laboratory scientists have developed ways of analysing hormones that ethologists can use to monitor reproductive condition and stress in free-ranging animals, by measuring hormone concentrations in fresh dung samples.

Because behaviour is so very variable, and so sensitive to interactions with the environment, it comes as no surprise that the behaviour of the animals of a single species can show quite marked differences from place to place. Over most of their range elephants drink at least once a day, and when feeding they routinely destroy large numbers of trees, setting up a cycle of alternation between woodland and grassland that can dramatically lower biodiversity locally, but that increases it over larger areas. However, in the Kaokoveld of Namibia, an area of sparse and erratic rainfall where vegetation and water are widely scattered across a bare, rocky landscape, elephants can go without drinking for three to four days, and when feeding in the groves of acacia trees along drainage lines they do not demolish the trees.

Ethology is essentially about wild animals in their natural habitats, and those wild places are diminishing day by day. Demands for living space from relentlessly rising human populations continually reduce the areas in which natural ecosystem processes can still occur. Africa is now the only continent that still has large numbers of big animals in natural habitats, but even in Africa migrations are constrained by fences and by human settlement of former wildlife areas. Only in the largest of protected wildlife areas is it still possible to witness the interactions among the large carnivores and herbivores. As well as tenaciously conserving what natural habitats are left, we must take while we can the opportunities that remain for field work on ecology and ethology, not least because the results of that work will prove essential in the fight to save wild places for the future.

SENSES

Animals continually gather information about their surroundings, and each animal's senses of sight, hearing, smell, taste and touch are adapted to its habitat and lifestyle. For example, a mole-rat, which spends almost all its time in pitch-dark tunnels, has tiny eyes and very poor vision, but at the tip of its tail it has a brush of fine but stiff bristles that are exquisitely sensitive to air currents and vibrations. Their function is to detect the draughts that warn of breaches in the mole-rat's burrow systems, and the vibrations that other mole-rats create by drumming on the sides of their burrows. Golden moles are even more specialised for life underground: their ears are covered by fur, and their eyes are completely covered with skin. In contrast, cats and bushbabies that hunt nocturnal prey have large eyes that give them excellent night vision, and large mobile ears for acute hearing. Springhares' large eyes and ears were designed through natural selection to detect the stealthy approach of nocturnal predators with similarly acute senses.

To a human observer, antelope seem to be in a state of constant nervous tension, continually scanning for danger and startling for no reason. In reality, their apparent nervousness is responses to stimuli that are too weak for human perception, but that the acute senses of the antelope pick up clearly. Non-human mammals have senses so finely tuned that they can communicate, navigate, forage and avoid enemies by sensory signals that humans cannot even detect. With these abilities they have no need of telepathy, extra-sensory perception and mysterious 'sixth senses', and there is not a shred of evidence that they have them. In fact, it demeans the animals to explain their abilities by mystical mumbo-jumbo just because we cannot credit their true abilities.

SIGHT

Cats can see in light so faint that humans call it darkness. Bushbabies have night vision even more acute, provided by eyes that are so large they do not move in their sockets, and by starlight a bushbaby can leap two metres between branches and not miss its footing. A suricate sentinel (*see* Chapter 5) can distinguish a harmless vulture from a dangerous eagle when both are just specks against the midday sky. Antelope, nocturnal predators, bushbabies, rabbits, hares and springhares have eyes that shine if a bright light falls on them at night. This glow is the reflection from the tapetum,

a layer of shiny crystals that lies just behind the retina at the back of the eye, and which bounces incoming light back through the retina, increasing by 60 per cent the chances that it will trigger a nerve impulse and contribute to the picture that forms in the animal's brain.

There is no connection between the time of day or night that a carnivore is active, and whether or not it has dark markings on its face. A diurnal cheetah's tear stripes have the same function as a nocturnal civet's bandit mask: both emphasise facial expressions that are used as social signals.

The deadly accuracy of a hunter's final pounce and the astonishing acrobatics of baboons, monkeys and bushbabies need precise perception of distance. Predators that need to judge the speed and distance of their fleeing prey, and primates that must judge the distance to the next branch in their leaps through the trees, have forward-facing eyes, with a wide area in which the visual fields of the eyes overlap each other. Anything in that area of binocular vision can be seen by both eyes, and the brain's unconscious processing of the image pair measures the distance to the object.

Above and below: The striking black patterns on a civet's or a cheetah's face emphasise their facial expressions.

Opposite: Forward-facing eyes provide a cheetah with the precise judgement of distance that successful hunting demands. Cheetahs, like other carnivores, see the world in pastel shades.

To its prey, the simple fact that a predator is there at all is of far more significance than a fine estimate of exactly how far away it is. Therefore, animals that are hunted rather than hunters, from small rodents and insectivores all the way up the size scale to rhino and hippo, need a visual field that is as wide as possible. To provide it, their eyes face sideways and have fields of view in which approaching danger can be spotted over nearly a half circle on each side, leaving only a narrow blind spot directly behind. The field of binocular vision is restricted to a narrow wedge directly in front, but this is of no consequence to animals that are not acrobatic climbers and whose food does not flee for its life.

There can be two types of light-sensitive cells in the retina of a mammal's eye – tall, narrow rods that are sensitive to the whole spectrum of visible light, and tapering cones whose peaks of sensitivity lie at three different points on the spectrum. These three peaks of sensitivity, for blue, green and red light, provide the basis for colour vision. Because they are essential for colour vision, examining an animal's retina for the presence of cone receptors tells us whether it could see in colour or not; if there are no cones, vision can only be in shades of grey.

Nature has no perfect solutions, so there is a trade-off between colour vision and night vision. Because the colour-blind rods are both smaller and more sensitive than the cones, a retina with many rods and few cones provides a clearer image in dim light than one with more cones. Primates and squirrels, active during the day and feeding on fruit, have colour vision so that they can tell ripe from unripe – the colour of a ripe fruit signals that it is rich in nourishing sugars, and that the bitter, toxic alkaloids and tannins that it contained while it was green have disappeared. Nearly all rodents, which make up 40 per cent of mammal species,

Top: With widely overlapping fields of view a primate's forward-facing eyes are adaptations to life in the trees.

Above: Sideways-facing eyes allow this impala to spot the approach of danger from almost any angle.

Right: Only mammals have large, mobile external ears.

Opposite: A leopard's whiskers detect vibrations and air currents as well as contact with solid objects.

are active at night when light levels are too low for colour vision. They have little need for the ability to see in colour, but they do need to detect the shifts of shadow and darkness that might warn of a predator's stealthy approach. Consequently their retinas are packed with very sensitive, but colour-blind, rod receptors.

Midway between diurnal primates and nocturnal rodents lie the carnivores, with enough retinal cones to see the world in pastel shades. Shape and form are far more important to a predator than colour; prey moves fast and erratically and certainly does not ripen like fruit! Perhaps what little colour vision predators do have helps them to avoid brightly coloured insects that are distasteful, poisonous or venomous.

HEARING

What are usually referred to as ears are actually only the mobile external part of a mammal's hearing apparatus. The actual sense organ is a fluid-filled spiral buried deep within the bones of the skull. Sound reaches it via the external ear canal, the eardrum and a chain of three tiny bones which evolution has moulded from parts of the jaw bones of mammals' reptile ancestors. The external ears, known as pinnae, scoop vibrations out of the air and funnel them down the external ear canal, enhancing the detectability of very weak sounds. In many species the ear pinnae have taken on other functions – their movements signal a mammal's moods, and in hares and elephants they are richly supplied with blood, which cools as it passes through them.

There is a trend for smaller mammals to be able to hear sounds higher up the frequency scale. Dogs can hear up to 40 000 Hertz (a Hertz is one vibration per second), cats 78 000 Hz, rats and mice up to 100 000 Hz, and some bats 210 000 Hz. Elephants can hear 10 Hz, a frequency below the range of human hearing, which they use for long-range communication (*see* Chapter 6).

Because they can swivel their external ears, most mammals are better at locating the source of a sound than humans are. A lesser bushbaby's huge, mobile pinnae allow it to locate a buzzing insect accurately enough for it to be snatched out of the air in the dark. A bat-eared fox locates buried dung beetle larvae from the sounds of their feeding, and can pinpoint them so accurately that it can dig directly down to them. The rumble of distant thunder can send wildebeest trekking towards where rain may have stimulated the growth of new grass.

TOUCH

Quite apart from its other unique properties, the whole of a mammal's skin is a sense organ, with sensory nerve endings that respond to temperature and pain as well as touch.

Rock dassies and yellow-spot dassies, which take refuge in narrow rock crevices, have long, black, touch-sensitive hairs sprinkled thickly throughout their coats, while tree dassies, which nest in tree holes, manage without. A Cape clawless otter locates its crab and octopus prey by feeling for them with its dextrous forepaws.

The sense of touch merges with hearing. A carnivore's whiskers, called vibrissae, can detect tiny air currents as well as contact with solid objects, and a fur seal captures fish in the dark by homing in with its long, stiff vibrissae on the vibrations that the fish make as they swim. A mole-rat has a tuft of fine bristles at the end of its tail, which help to detect vibrations in the soil through which it burrows.

SMELL AND TASTE

For the majority of mammals the sense of smell is overwhelmingly the most important. They live in a world of richly subtle odours that humans cannot even conceive of. Even the concentrations at which mammals can detect odours are difficult to comprehend – one part in a thousand million million, equivalent to one drop in 10 million tons of water. That sensitivity, and a specificity that allows discrimination between hundreds of thousands of different odours, enables mammals to use their sense of smell to find and select food and mates, navigate, communicate and avoid danger. Some mammals respond to substances that humans cannot smell at all, even at high concentration, and domestic cats and dogs and their wild relatives seem to be able to taste water itself.

The olfactory epithelium, which is where scents are actually detected, is a small patch of yellowish tissue near the back of the nasal cavity. Its sensory cells are connected directly to the brain, whereas signals from all the other sense organs must pass through at least two synapses.

A fruit bat selects the ripest fruit from a bunch on the basis of smell. Insectivores, invertebrate-eating small carnivores and pigs are guided to buried food by their noses. Scavengers such as jackals, hyaenas and bushpigs locate carrion by its odour – hyaenas can pick up the aroma of a ripening carcass lying as far as 4 km upwind.

Once a mammal has followed its nose to food, its sense of taste is the final arbiter of whether it is fit to eat. Sensations of sweet and savoury signal nutritious soluble sugars and proteins respectively; sour, acid, bitter and hot warn of toxins. Small carnivores are repelled by the poisonous and foul-tasting secretions of toads, millipedes and grasshoppers, and the number of termites that an aardwolf can eat is determined by its tolerance for the taste of their defensive secretions.

Insectivores, bats, elephants, ungulates, carnivores and some primates have one sense more than man. In the roof of the mouth is the Jacobsen's organ, also called the vomeronasal organ, which seems to respond to both 'smells' and 'tastes' in a way that has no parallel in human experience. The Jacobsen's organ connects to the nasal cavity or the mouth, or both, by paired narrow ducts, and air- or liquid-borne chemicals are pumped into it by muscular contractions whose outward signs are the curling of the upper lip and wrinkling of the nostrils that go by the name of flehmen. Jacobsen's organ is specially designed to detect steroid hormones and similar substances, and flehmen is most often seen when a male is testing a female's reproductive condition (*see* Chapter 7), though in herds of eland and gemsbok females also use flehmen to test one another's urine.

LEARNING

Some ability to learn, to modify behaviour as a result of experience, is almost universal among animals – even worms can learn not to respond to a stimulus that is repeated over and over again. Even so mundane an activity as returning to a nest demands that its location, or the path back to it, be learned and remembered. Most animals have rather limited learning abilities, although a few species are 'experts' at very specific tasks (like the solitary wasp that recalls the location of its nest and how much food it has stored there), but birds and mammals stand out by virtue of the wide range of what they can learn. Of the two groups, mammals show the greatest learning abilities, mainly because there are no bird equivalents of primates. Crows and parrots are probably about as clever as dogs and dolphins.

Top: A ram uses the flehmen grimace to detect steroids in a female impala's urine.

Above: A dik-dik sniffing a blade of grass. For mammals the sense of smell is overwhelmingly the most important.

Above and left: Male mammals use flehmen to check the reproductive condition of females.

The ability to learn shows some puzzling and unexpected differences. Ants learn mazes as quickly as rats, and when food is the reward rats are better at learning pattern discriminations than chimpanzees or human two-year-olds. The question of which components of an animal's repertoire are learnt and which are innate also shows some puzzling patterns. From an age of about four months, young dwarf mongooses of both sexes join sentinels while they are on duty (*see* Chapter 5) and learn how to guard, but it is usually only the males that later become sentinels themselves. A group's battle formations in attacks on predators and conflicts with other mongoose groups do not depend on learning – they are the result of innate differences in the behaviour of mongooses of different ages.

Maternal care frees young mammals from the harsh necessities of independent living and offers them the opportunity to grow, develop and learn. Below a baby baboon practises the manipulation skills it will need as an adult.

Mammals do most of their learning while they are still young, when the care that their mothers lavish on them gives them time to experiment and allows them the luxury of making mistakes (*see* Chapter 2). For a baby elephant, learning to control and co-ordinate its trunk's 5 000 pairs of muscles is, pardon the pun, a mammoth task, and it takes about three months to transform a trunk from a floppy appendage to a useful tool.

Avoiding brightly coloured insects that sting, taste foul or are poisonous demands that the association between colour and unpleasant consequences be learned. Opportunistic omnivores such as rats and jackals very readily learn associations between food flavours and illness, and if a food has once made them ill they never touch it again. If the food is a poison bait and they eat less than is needed to kill them but enough to make them sick, they will subsequently avoid the bait. Rats can learn by sniffing noses what others have been eating, and black-backed jackals seem to be able to learn by watching other jackals – if a pup sees its mother avoid a poisoned bait or trap it will learn to do the same. This cultural transmission of food preferences and aversions can make whole populations of mammal pests impossible to control by baits and traps. Rodent control is an arms race between bait design and rodents' learning. Where small stock are farmed, trap- and bait-wise jackals are a major problem, which is only made worse by the traditional, heavy-handed blanket control measures that kill off their less discriminating competitors.

Lions can learn preferences for certain foods, and this can lead to prides specialising in hunting particular species of animal. One pride in the Kruger National Park specialises in porcupines, and another in Selinda, Botswana, in hippos. In Savuti, Botswana, an unusually large pride of unusually large lions have turned their physical prowess to good effect by learning to kill young adult elephant bulls. Food preferences can pass from adults to young, and thus be perpetuated for longer than the lifetime of a single animal – a culture of man-eating in the lions of the Njombe area of Tanzania lasted for 15 years, and could well have gone on indefinitely had the lions not been hunted down and killed.

Practice at capturing and killing a particular type of prey will enhance success in later hunts – predators do better when they concentrate their efforts on those species that they are best at. Any special techniques that enhance success will be learned and passed on in parallel with the preference for a particular prey, one cycle of learning reinforcing the other. Some wild dog packs immobilise large prey using a bite to the nose, which has the same paralysing effect as a twitch on a horse. Although use of the technique occurs in both Hwange and the Okavango, and did occur in the Serengeti until the wild dogs there were wiped out by rabies and canine distemper, not all packs use it, and it is apparently passed on from adults to pups. It would be interesting indeed to know whether founder groups could introduce the technique to their new packs. Probably they could: when captive-bred females and wild-caught male dogs were combined into a new pack at Madikwe Nature Reserve, South Africa, the females quickly learned hunting techniques from the males.

Animals such as hyaenas and squirrels that store surplus food in scattered caches have to remember not only that they made a cache but also where it is. When their spatial memory can be tested they show an astonishing ability to relocate caches made days or weeks before.

Learning is especially important for members of social groups, who have to remember the identities and status of other animals in their group and how they behave in friendly and competitive interactions, in order to keep track of changes in

Below, left: Until they learn to use their trunks, baby elephants drink directly with their mouths.

Below and bottom: An adult elephant's trunk is a precision instrument, used to pluck a single leaf from a twig or to skim water from a muddy pool.

group composition and dynamics and the shifting web of social structure. Laboratory experiments show that close relatives, particularly family members, can recognise that they are related even when they have been raised separately, but in the wild there is no question that parents' recognition of their young, and siblings' recognition of each other, are based on learning.

Territorial animals have to learn from other animals in their group where their boundaries are, even if these are posted with scent marks. Where a territory is held by a group such as spotted hyaenas or suricates, for example, the youngsters learn where the boundaries are by accompanying the adults on border patrols and in battles with neighbouring groups of animals. In a very real sense they inherit the territory. Among brown hyaenas on the southern Namib coast, cultural inheritance

In National Parks and nature reserves encounters like this with habituated animals are becoming routine.

of territories has led to a tradition of visiting long-abandoned human settlements inland where their predecessors scavenged garbage, even though today their only food sources are found along the coast.

The success of man-made territorial borders – in the form of electric fences – at containing large animals like elephants depends on their learning that the fence is dangerous. Animals that are translocated from areas without electric fences are given a period of boma training in which an electric fence is backed up by a sturdy palisade of poles and cables, which teaches them that electric fences bite, but denies them the opportunity to learn that a high-speed charge can break them down.

The viability of ecotourism, with all that it means for sustainable conservation of wildlife areas, depends heavily on animal learning. Modern game viewing would be impossible if animals in wildlife areas had not learned, through habituation by continual exposure, that vehicles are harmless. It is now possible to get closer than any of the famous white hunters – closer even than Khoisan trackers – to animals behaving naturally in their natural habitat, simply by driving along the roads of national parks or nature reserves. One pair of young lions in the former Kalahari Gemsbok Park became so habituated to vehicles as they grew up playing among the vehicles at the Cubitje Quap waterhole that they invented a game of biting holes in tyres!

Opportunistic omnivores like this baboon quickly learn to exploit novel food sources.

A long and growing list of mammals – banded mongooses, ratels, genets, elephants, ground squirrels, lesser bushbabies, woodland dormice, tree squirrels, baboons, vervet monkeys and warthogs – have all learned that rest camps in wildlife areas offer food and safety from predators, and that the people in them are largely harmless, or actively beneficent. Baboons and vervet monkeys very quickly learn that humans are a potential source of food, and learn from each other how to open rubbish bins. Where their raiding habits and aggression against people mean that they have to be shot, they learn to fear people in Parks Board uniforms, and even the approach of official vehicles. It is a pity that visitors to wildlife areas cannot learn not to feed monkeys and baboons.

Recently, reports have surfaced about banded mongooses grooming warthogs at safari lodges where animals of both species have become tame. Similar behaviour has never been seen in the wild: it very probably arose as a consequence of repeated interactions between the same individuals of the two species, with their meetings being much more frequent than they would be in the wild because abundant, readily available food and the absence of predators had reduced their need to move around. If the inter-specific grooming has arisen only in the 20 years or less that safari lodges have been operating it must be a learned behaviour, because that time span is much too short for the evolution of a new behaviour by natural selection.

CHAPTER TWO

BIRTH AND INFANCY

beginnings and becomings

GIVING BIRTH

LACTATION

HIDERS, FOLLOWERS AND RIDERS

LEARNING AND PLAY

CARNIVORE YOUNG – BORN KILLERS

Previous pages: *Once an impala lamb can follow its mother it joins the breeding herd. At first the mother and lamb stay close together but when the lamb is fully mobile it mixes freely with the herd and only goes back to its mother to be suckled.*

Below: *Birth – for the newborn springbok the beginning of individual existence, for the mother merely the transition from gestation to lactation.*

ALTHOUGH WE DO IT OURSELVES, and therefore take it completely for granted, being born is actually a rare and special event in the animal kingdom. Unlike mammals, the overwhelming majority of animals begin their active lives by hatching. Apart from a handful of species of cold-blooded vertebrates and a few invertebrates, mammals stand alone in that they begin their development inside a mother's body and only leave it for the outside world when all their developmental building blocks are already in place.

In all Africa's mammals there is an intimate physical connection between the developing foetus and the wall of its mother's uterus. This connection, the placenta, through which oxygen and nutrients are passed to the foetus and its waste products are carried away, leaves the foetus with nothing to do except develop and grow – it is warmed, protected and fed directly by its mother's body. The development of mammals' large brains, with all the potential for complex behaviour that they provide, would be impossible without the steady stream of energy and nutrients that the placenta transfers to the foetus. And birth is only a comma in the sentence of a young mammal's dependence on its mother: she continues to feed it by lactation, grooms it

and protects it, and in many species she helps it to learn the survival skills that it will need in adult life.

Mammalian reproduction and lactation are intriguingly diverse. Rabbits give birth to helpless, naked, blind and deaf young, while newborn hares are fully furred, can see and hear, and are fully mobile after two days. In both rabbits and hares parental care is limited to a single suckling session less than ten minutes long in each 24 hours. Elephant shrews, whose closest relatives may be rabbits and hares, are born fully furred and completely mobile, with all their senses fully functional; they are weaned after only about a week, and are independent within a month. Spotted hyaenas are suckled for up to 18 months, and elephants for up to eight years, and social bonds between mothers and daughters are life-long.

The feeding of the young with milk produced by the mother is a defining characteristic of mammals.

GIVING BIRTH

During labour and delivery the mother is terribly vulnerable to predators, and in nearly all species of mammal females give birth in secure dens, or in the cover of dense vegetation, which also serve as havens for their newborn young. (Blesbok and black wildebeest are two of the few species in which females do not leave their herd to give birth.) A warthog mother modifies a burrow so that her litter can rest in a raised recess, out of the reach of water that might run into the hole. A female bushpig gathers a heap of grass about a metre deep, and then burrows into it to form a refuge where she gives birth and keeps her young. Giraffes have calving grounds, where the cows go to give birth even if their usual home ranges are some distance away. Apart from whales and dolphins, hippos are the only African mammals that ever give birth in water. Usually the water is shallow enough for the baby to breathe while it stands on the bottom; only very rarely does it have to swim up to the surface to take its first breath.

Even in those few species where the males help to raise the young (*see* Chapter 7), they play no part in the birth process itself. Elephant mothers have other females as attendants when they give birth – but no one knows what happens when a dwarf mongoose or suricate female gives birth in a den where the rest of the social group are also present.

Immediately a baby is born its mother begins to vigorously lick and nuzzle it, removing the birth membranes and drying its fur. She bites through the umbilical cord, which has connected the developing foetus to its placenta throughout gestation, and – especially if the young are to stay in the den where they were born – she eats the placenta and licks up spilled birth fluids to keep the nest clean and to eliminate odours that might attract predators.

For a carnivore mother the placentas of her litter are a significant source of nourishment, which will carry her through the fasting of the first day or two when she will not leave her young for long enough to hunt. For a herbivore mother the licking and cleaning of her young demand a marked change in her behaviour, because the smell of birth fluids is usually repulsive to animals that do not eat meat. However, the passage of the baby through the birth canal triggers chemical changes in her brain that make the smell of birth fluids temporarily attractive.

In all species the mother's cleaning of her newborn young fixes their odour in her memory. Once that has occurred, and the chemical changes in her brain have dwindled, she will not accept any but her own offspring. The period during which a female will accept other young varies from species to species – in antelope it is very short, a few hours at most; while in carnivores, whose young are confined to a nest or den where the chances of the female encountering young from other families are small, it extends until the young begin to leave the nest. Many carnivore mothers move their young to a new den every so often, probably to avoid a build-up of scent that might attract predators, but also perhaps to escape from blood-sucking external parasites such as lice and fleas.

A zebra mother returns to the safety of her breeding group as soon as her foal can walk well enough to follow her, but for its first two or three days she keeps between it and the rest of the group so that it learns only her stripe pattern – zebra mares are intolerant of approaches by any but their own foals, and use hooves and teeth to drive others away, sometimes inflicting disabling or fatal injuries. Blue wildebeest and eland cows, and female Cape fur seals, are similarly intolerant of any but their own offspring, and an unweaned baby that loses its mother is doomed, because no other female will adopt it.

Opposite: In its first two days a zebra foal learns its mother's stripe pattern.

Below and bottom: Young bat-eared foxes and cheetahs are provided with milk and solid food until they are able to capture their own prey.

LACTATION

The feeding of the young with milk produced by the mother is a defining characteristic of mammals, and its ramifications spread through the whole of mammalian biology. Because they lactate, most female mammals are able to raise their young without assistance from a male, and so only about 10 per cent of mammalian genera have members whose males help care for the young, and in only about five per cent of mammal species is the mating system monogamous (*see* Chapter 7).

Mammary glands are modified apocrine sweat glands: teats, nipples, udders and breasts are simply ways of delivering milk to the young. The precise composition of milk varies dramatically between species: zebra milk contains almost no fat, but is very rich in sugars, while to ensure that her offspring grows as quickly as possible a female seal produces milk that contains 50 per cent fat. Newborn young are fed on a special milk called colostrum, rich in antibodies that protect them from disease until their own immune systems have matured.

Being fed on milk allows young mammals to delay their development of features that would be necessary for foraging independently – such as a full set of functional teeth, which develops only after weaning, when the skull is much closer to its adult size. Because the permanent teeth grow in jaws that are nearly fully formed they can develop a precision fit that provides very efficient processing of food – a lion's jaws close with the smooth snick of precision machinery because lion cubs do not grow even their milk teeth until they are a month old, and are nearly two years old by the time they have their full adult dentition.

Another consequence of lactation is that close social ties between mother and young and between litter siblings are maintained at least until weaning – and these early relationships are often carried forward into the lifelong alliances between family members that form the basis of the most complex and stable of animal social groups (*see* Chapter 5).

Although they are being fed on milk, young antelope and zebra start nibbling at vegetation when they are only a few days old, and baby dassies can eat solid food after only one day. This early introduction to their adult diet is probably necessary for the growth in their guts of the microbes which break down the cellulose in the plants that they eat. Baby carnivores live on nothing but milk for at least a few weeks, before their mother or her helpers begin to bring them solid food. They have no need to kick-start a flora of intestinal bacteria because meat and insects are easier to digest than plants, and they have not yet developed the skill, strength and speed that they need to catch their own prey.

Top: At just over a year old this weaning giraffe calf takes one of its last drinks of its mother's milk.

Above: A baby elephant feels with its trunk for the nipple just behind its mother's front leg, but it sucks only with its mouth.

HIDERS, FOLLOWERS AND RIDERS

Because they are so much easier to catch and kill than the adults, young antelope are a favourite prey of carnivores from black-backed jackals upwards. To avoid being killed and eaten they adopt one of two strikingly different strategies: either they avoid detection by lying hidden in cover, or they move with the herd from an early age. Blue wildebeest calves are on their feet within a minute; they can run within five minutes, and after less than a day they can keep up with the adults. A grey duiker lamb, which is a hider, stands after 35 minutes and walks after 50. A sitatunga calf, born in the seclusion of a reed platform prepared by its mother in her permanent swamp habitat, can swim before it can walk properly. Even after a month a young klipspringer is still unsteady on its feet, and it has no choice but to lie tightly in cover until it has grown enough to match the boulder-hopping agility of its parents. While a calf or lamb is lying out its mother visits from one to four times a day, depending on its age and species, to suckle it. She eats its urine and faeces so that predators cannot scent them.

Like the rest of its behaviour, whether an antelope calf is a hider or a follower is an adaptation to its environment. Out on the short grass plains where blue wildebeest, black wildebeest, tsessebe and red hartebeest graze, there is no cover for calves to hide in, and so they are followers. Kudu, duiker, roan, sable, gemsbok, klipspringer and other species that live where long grass, bushes or rocks offer secure hiding places have hider young. Springbok and impala lambs hide for the first two days, until they are steady on their feet, and then join their mothers in the herds, although it takes a young springbok a month before it can match the acceleration of an adult. For its first 3–6 weeks a gemsbok calf stays in hiding during the day while its mother grazes up to 2 km away. At night it may follow its mother, and then find itself a new hiding place in the morning. By the time it begins to follow its mother during the day its horns have already begun to grow, which gave rise to the myth that gemsbok are born with horns.

A roan cow leaves her herd a few days before she gives birth to find a secure place for her calf to hide. For the first few days she stays on guard near her calf, then returns to the herd and for the next six weeks visits the calf morning and evening to suckle and clean it. After each visit the calf goes alone to a new hiding place so that predators cannot find it by following its mother's scent trail. When artificial water-holes in the north of the Kruger National Park allowed zebra and blue wildebeest to build up in numbers they grazed down the grass that roan antelope calves depend on for cover. The resulting heavy lion predation on the calves brought the roan population to the brink of extinction. Restricting the access of zebra and wildebeest to some areas by closing down water-holes has allowed the grass – and the roan population – to begin to recover.

To keep them under their protection, some mothers carry their young with them as they move around in search of food. A bushbaby mother stays in her nest continuously for the first three days after giving birth, and then carries her offspring with her when she goes foraging, leaving it clinging to a twig while she feeds nearby. A baboon mother uses one arm to support her newborn

An unseasonably late fire has robbed these impala lambs of the cover in which they would usually lie hidden.

baby as it clings beneath her chest. After a few hours it is able to grip her fur and hang on by itself. At five weeks, when it is old enough to walk, the youngster rides on its mother's back – at first crossways and face down, and later sitting up and using the mother's raised tail as a backrest.

Young Brants' whistling rats have a gap between their incisor teeth that hooks behind the tips of their mother's nipples. They cling so tightly that she can drag them from nest to nest in order to avoid predators. Young forest shrews cling to their mother's nipples for their first 5–6 days, then start clinging to her fur with their teeth, first in a cluster and by the thirteenth day in a chain, with each youngster gripping the fur of the one in front, a habit known as caravanning. Litters of greater and lesser red musk shrews also caravan, but they do not nipple cling, and if a mother wants to move her offspring she has to carry them in her mouth.

LEARNING AND PLAY

Lactation and parental care allow young mammals the freedom and the spare time to learn more during their early lives than they do – or need to – once they are adult. In all sorts of mammals, food preferences are influenced by youngsters' experience of what adults – especially their mothers – eat. A young zebra or antelope sniffs at its mother's muzzle as she grazes; a young vervet monkey watches intently as its mother feeds, and samples what she eats. As well as offering a chance for them to hone their capture and killing techniques, the prey that a female cat brings to her kittens or cubs also teaches them which species are appropriate as prey. When attempts are made to release captive-bred carnivores to the wild, their lack of early exposure to prey can lead to problems: with an enthusiasm born of naivety cheetahs released into Suikerkop Nature Reserve, South Africa, successfully hunted young giraffe.

In baboon troops the youngsters are the ones that learn most readily about new foods, since they are much more likely than their elders to try something novel. The other members of the troops then learn by observation and imitation.

From an age of about four months, young dwarf mongooses of both sexes join sentinels while they are on duty (*see* Chapter 5) and learn how to guard, but it is usually only the males that later become sentinels themselves. Baby vervet monkeys learn which of the five alarm calls is appropriate for which predators by observation of older monkeys.

Opposite and above: Mammals are the most playful of animals, but play's exact function is still uncertain.

The most playful animals are mammals, and the most playful mammals are young primates and carnivores. Although play is very easy to recognise, it is infuriatingly difficult to define in a way that distinguishes it from 'serious' behaviour. Not even the experts in the field can come up with a set of characteristics that can be reliably used to classify whether behaviour is playful or not. Exaggerated movements, rapid alternations of behaviour and role, repetition and incomplete behaviour patterns are unquestionably features of play, but in the end play is most easily and most reliably classified as behaviour that looks playful. That play is so easy to recognise is itself an adaptation; selected to ensure that play is recognised for what it is, so that partners can respond playfully.

Play is very much more common among mammals than among other sorts of animals. Among birds, only parrots, tits and crows and their relatives can definitely be said to play. No reptile has ever been seen playing in the wild, and there are only very few isolated records from captivity. If amphibians or fish play it is not in a way that we mammals can recognise.

Young hippos' playful sparring may be training for the bloody battles they will fight as adults against both rivals and predators.

Young mammals play far more than adults do. Young cheetahs in East Africa play 3–4 per cent of the time, and other species up to 10 per cent of the time. Play uses up 5–20 per cent of spare energy, and there must be some risk of injury during its more exuberant moments, plus increased risk of predation. With such an investment of time, energy and risk, what is the benefit of play, what is it for? Answering the third of ethology's 'Why?s' (*see* Chapter 1) about play has proved to be as difficult, and as controversial, as defining play in the first place.

The fact that young mammals are the most playful animals suggests that play is related to learning, because mammals' behaviour is more dependent on learning than that of other animals, and most of that learning happens before sexual maturity. The sort of play young mammals indulge in fits their adult lifestyles. Young springbok run races on dry Kalahari river beds, wild dog pups hunt their littermates, lion cubs pounce at their mother's twitching tail tip, and elephant calves mock charge each other and animals of other species. This correspondence between play and adult behaviour makes it look as if play is practice for future needs, but animals that grow up without playing, either in a laboratory experiment or perhaps because they were hard put simply to survive during a drought in the wild, grow up into capable adults – so if play is just practice, it is practice that animals can manage without. There are also significant differences in the details of play and serious behaviour that challenge the idea that play is practice. And play is not a very effective way of building physical fitness – it does not go on for long enough to enhance cardiovascular function.

Other suggestions for the function of play include social integration, building confidence for attacks on prey or rivals, or developing connections between nerve cells in the brain. Something as complex, common and widespread as play probably has a multitude of causes, effects and benefits.

In carnivores, play fighting and mock predation are the most common sort of social play, and there is an ever-present danger that if what is meant in play is taken in earnest, serious injuries could result. As a result a carnivore that wants to play with one of its group mates signals its intentions very clearly. Dogs of all species signal playfulness with a stereotyped play bow – raised rump, forelegs stretched out in front, raised tail wagging, an open-mouthed grin and eyes partially closed.

Bat-eared foxes are among the few animals that still play when they are adult: partners chase each other, they play with objects such as feathers and sticks, and

Playing as practice for preying. The sort of play that young animals indulge in fits their adult lifestyles; wild dog pups hunt their littermates in preparation for when they will live by capturing real prey.

they have even been seen playing chase with antelope in East Africa. To show that this was not an aberrant behaviour by a single individual, the observations were repeated in South Africa's West Coast National Park. Paradoxically, bat-eared foxes are among the least predatory of carnivores – their diet of termites, dung beetle grubs, grasshoppers and fruit is simply collected rather than hunted. Perhaps play provides an outlet for unexpressed predatory behaviour.

CARNIVORE YOUNG – BORN KILLERS

LIONS

After a gestation of 14–15 weeks a lioness leaves her pride to give birth to a litter of up to six cubs in secluded heavy cover. The cubs weigh only 1,5 kg each, just one per cent of their mother's body weight. Although she may leave the cubs alone for up to 24 hours, the need to suckle and care for them restricts the distance over which she can move, and if her pride moves away she has to hunt alone, which can make her food supply more erratic and less predictable. She does not introduce the cubs to the pride until they are 6–8 weeks old (or older if there are other, bigger cubs already present that would compete with hers for milk).

Opposite: Young predators are born with a hunting instinct, but the finer points of capture and killing are honed during practice sessions with dead and live prey brought to them by their mother.

Below: A lioness keeps her cubs away from her pride for their first 6–8 weeks.

Once they have been introduced to the pride the cubs are suckled by all the females that are in milk, with each lioness showing only a weak bias towards her own offspring. It is not unusual to see a lioness suckling cubs of widely different sizes, but she will terminate a suckling bout when her own cubs are satisfied, even if others want to continue. This allomothering evolved by kin selection (*see* Chapter 1); because the lionesses in a pride are genetically related, on average as closely as second cousins, they all share a genetic stake in the survival of the cubs.

Despite the lionesses' communal care and their fathers' indulgent tolerance, up to nine out of ten of a lion population's cubs never see their first birthday. Quite apart from infanticide during pride takeovers (*see* Chapter 5) they also suffer seriously from starvation during lean periods because they are quite unable to compete with the adults in the scramble for a share of a kill.

HYAENAS

The implacable force of natural selection has made giving birth a remarkably trouble-free process among African mammals, with one striking exception: spotted hyaena females giving birth for the first time suffer a nine per cent mortality rate and 60 per cent of their young are still-born. The reason is the aberrant anatomy of the females' reproductive tract: because female foetuses are exposed to levels of male hormones as high as those in adult males their genitals become masculinised (*see* Chapter 5) and the birth canal not only has a right-angle bend about half way down, it is also four times as long as the foetus's umbilical cord. For three-quarters of its journey into the world a newborn spotted hyaena is cut off from any supply of oxygen. The occurrence and persistence of such a situation is due to the compensating benefits that being masculinised confers on female hyaenas when they are adult: they are larger and more aggressive than males, and the largest and most aggressive of them raise two and a half times as many offspring as any of the rest because they dominate others in competition for food and secure den sites.

Spotted hyaenas nearly always give birth to twins. The newborn cubs' canine and incisor teeth are fully erupted and they begin life with a savage fight for dominance. If both of the twins are female, that fight is invariably fatal for one of them – both twins survive only if both are males or the sexes are mixed. As with the spotted hyaena's perilous birth process, the selective force driving this siblicidal rivalry is the increased reproductive success of dominant females. By killing her sister a female cub monopolises her mother's milk, which gives her a head start in the race to become bigger and more powerful than the other females in the clan. She also avoids having to compete with her sister later over their inheritance of their mother's social rank.

Spotted hyaena cubs are not weaned until they are 15–18 months old.

Spotted hyaena cubs are suckled until they are 15–18 months old, when they can follow their clan on its hunting trips. Until they are a year old they cannot survive without their mothers. This reliance on milk, rather than on meat brought back to the den, has two origins. Sometimes the mother is away hunting or foraging for as much as two days – and converting meat into milk is an effective way of keeping it fresh. And milk cannot be stolen by lions or other hyaenas as she returns to her cubs, by dominant adults when she reaches the clan den, or from her cubs by other cubs.

Brown hyaena cubs are not completely weaned until they are 12–16 months old, but in sharp contrast to their spotted relatives they start eating meat when they are as young as three months, provisioned by their mother's clan mates. Since raising cubs is a collaborative effort for a brown hyaena clan, there is little chance of meat being stolen at the den, and living in areas where large herbivores and therefore large predators are rare reduces the chance of having food stolen on the journey home. Those journeys can be very long, and the loads heavy; one brown hyaena in the Kalahari carried a 7,5 kg chunk of a carcass for 15 km.

WILD DOGS

A wild dog pack is nomadic, probably so that it does not have to hunt repeatedly in the same area, or in order to avoid the attentions of spotted hyaenas, which steal the dogs' kills, and of lions, which both rob the dogs and kill them. So that raising pups anchors a pack to its den for as short a time as possible, all the pups are born to the dominant female in a single litter up to 21 strong. Only very rarely does a second female in a pack give birth, and then the dominant bitch kills or forcibly adopts her pups. Within three weeks the pups are eating meat regurgitated for them by all the pack members, and from then on, under the care of the pack, they can survive the loss of their mother. At 5–10 weeks old they are weaned, and at 13 weeks they leave the den and follow the pack in its search for new hunting areas.

Like brown hyaenas, all members of a wild dog pack contribute to raising the pups, and even adults share meat, but instead of carrying meat back to the den in their mouths, which would leave it vulnerable to being stolen by other predators, they bolt it down at the kill, carry it home in their stomachs and then regurgitate it in response to begging by the pups or by the adults that stayed behind on babysitting duty. This behaviour has undoubtedly evolved from the basic canid habit of a mother regurgitating food for her litter, which jackals and Cape foxes do when they begin to wean their cubs, changing later to feeding them with food carried home in the mouth.

Wild dog pups are fed on regurgitated meat by adults returning from the hunt.

CHAPTER THREE

FOOD AND WATER

the bare necessities

FOOD

HERBIVORES

PREDATORS

BATS

WATER

FOOD

There are around 4 000 species of mammals in the world, but they can be divided into just two sorts: those that eat flesh and those that don't. 'Flesh' of course includes fish, reptiles, amphibians, insects and other invertebrate creepy-crawlies, as well as the meat of birds and mammals. Splitting mammals into those that eat plants and those that don't produces a hugely uneven division; even the most committed of carnivores will eat wild melons for the water that they contain. Probably it is only some marine mammals, and perhaps otters, that never eat plant material. Eating plants can provide just as varied a diet as being a predator: grass, broad-leaved herbs, and the leaves, seeds, bark, wood, roots, flowers, fruit and gum of trees are all eaten by mammals of one species or another.

With such a diversity of foods it is no surprise to find a corresponding variety of adaptations in feeding behaviour, and in the ways that food is found, harvested and consumed. Styles of feeding still label some of the orders of mammals, but as our knowledge of ecology and behaviour has expanded, the labels have begun to hang rather loosely. The insect-eating label of the order Insectivora sticks tightly enough to the shrews, hedgehogs and golden moles that are its members, but bats, aardvarks, aardwolves, pangolins, elephant shrews and bushbabies – in the orders Microchiroptera, Tubilidentata, Carnivora, Edentata, Macroscelidea and Primates, respectively – also live largely on insects. Carnivore means flesh-eater, which is accurate enough for dogs and cats, but members of the order Carnivora eat a vast range of other foods: small mongooses and bat-eared foxes eat insects, otters eat fish and crabs, and palm civets live on fruit.

Previous pages: Ground squirrel feeding.

Above: Giraffes do 90 per cent of their feeding above the reach of competitors.

Below: Unlikely carnivore, a chacma baboon feeding on a baby Egyptian goose.

At least partly as an evolutionary result of competition against other animals, some mammals have become specialists on particular sorts of food. Aardwolves have carved themselves an almost exclusive ecological niche by being the only mammals that eat snouted harvester termites. Pangolins have specialised on ants, and civets eat large numbers of millipedes that other mammals avoid because they have foul-tasting and toxic defensive secretions. Lions' collaborative hunting allows them to take prey that is too large and dangerous for any other carnivore to tackle. A giraffe's height allows it to do 90 per cent of its feeding above the reach of competing browsers such as kudu and eland; moreover, to reduce the extent to which they have to compete with each other, male and female giraffes employ slightly different feeding styles. To get full benefit from their extra height giraffe bulls feed at full stretch with their heads tipped back, reaching up to 5,8 metres, while cows avoid competition from bulls by feeding in a slightly lower band with their heads tipped downwards. Warthogs' tusks enable them to root in hard soil, allowing them to penetrate drier habitats than can be used by bushpigs which can only root in soft, moist soil.

Warthogs' tusks enable them to root in hard ground, and allow them to exploit drier habitats than bushpigs. The latter can only root in soft, moist soil.

HERBIVORES

The least selective of all the herbivores are elephants: they eat anything from roots to fruits, and when feeding on the foliage of trees they swallow as much wood as leaves. If the bunch of fresh leaves at the top of a tree is out of reach, they will push the tree down to get at it. They chisel bark off trees with their tusks, and peel it off in long strips with their trunks. They expose roots by digging with their forefeet, and then prise and pull them loose with their tusks and trunks. Tussocks of grass are pulled up and bashed against a front leg to dislodge the soil from the roots. Bull elephants shake full-grown camel-thorn trees to bring down a rain of tasty, nutritious pods, which cows and calves that are not strong enough to shake the trees take shameless advantage of.

An elephant's feeding strategy is to consume a huge amount of a wide variety of plants, skim off the easily digested nutrients, and then dump the rest. It chews its food only roughly, and passes it quickly through its cavernous gut, depositing 56 per cent of the 150–200 kg that it ate the previous day as coarse, fibrous dung balls. (That elephants get drunk because they have eaten fallen fruit that has fermented inside them is just another of the myths of Africa – the conditions in an elephant's stomach do not allow the sort of fermentation that produces alcohol.)

A black rhino crops twigs and shoots with its cheek teeth, leaving them snipped neatly at a 45° angle. Fragments of these twigs, with their tell-tale ends, are the most reliable way of distinguishing black rhino dung from that of white rhino, which contains only grass. Elephant dung also often has twigs and sticks in it, but with their ends torn and chewed off, not snipped neatly.

Tree rats cut the growing tips off the twigs of the Kalahari camel-thorn trees that they live in, and carry them back to their nests to eat. They add the uneaten leftovers to the nest, which soon overflows the tree hole in which it was built. On the ground, Brants' whistling rats also carry food back to their warrens to be eaten, and the area around an active burrow is littered with twigs and inedible debris. By eating green vegetation – and especially succulents – they get all the water that they need from their food.

Giant rats, pouched mice, Cape gerbils and Sloggett's rats carry food back to their burrows and store it rather than eating it immediately. Their larders allow them to take advantage of short-lived gluts of food, and provide a reserve that they can draw on if foul weather or danger from predators restricts their foraging. A pouched mouse burrow that was dug up was found to contain 700 ml of seeds and husks, and one giant rat's hoard contained 8 kg of macadamia nut shells. Cape mole-rats and common mole-rats store bulbs and corms near their nests – one common mole-rat collected 4 944 of them! Porcupines too carry food back to their dens, together with bones that they gnaw as a source of calcium and phosphorus and perhaps to sharpen their teeth. Bones lying around a porcupine burrow may make it look as if a hyaena is in residence, but the twin marks of the porcupine's 15 mm-wide incisors are a unique field sign of which species is responsible.

In their war against the depredations of herbivores, plants cannot run away and they cannot hide, but they can fight back, with an array of defensive spines and toxic and repellent chemicals. A giraffe feeding on a thorn tree is living proof that spines and hooks do not stop a browser from feeding. But they do slow it down – a giraffe cow has to feed for 12 hours a day to fulfil her dietary needs, and a bull for 9–10 hours. The benefit to the tree of having physical defences is that it loses fewer leaves than it would if it were not spiny.

Compared to animal food, plant foods are seriously lacking in nutrients, and those that they do contain are often difficult to digest. Although fruit is rich in sugars, it is lacking in protein, and all fruit-eating mammals, with the exception of fruit bats, have to supplement their diet with insects. Fruit bats have very rapid digestions that skim off what little protein their food contains and dump the excess sugars that are not needed to provide energy for flight. In any case, by the time it is ripe enough for a bat to eat, most wild fruit is riddled with insect larvae, which supplement the diet of any fruit-eater with a little fresh meat.

Nearly all the energy in plant foods is locked up in giant molecules of cellulose that have to be broken down before they can be assimilated, but not even the most herbivorous of mammals produces the digestive enzymes necessary to effect that

Opposite, top: Each day an elephant's non-selective bulk feeding turns 200 kg of vegetation into 100 kg of dung.

Opposite, bottom: The sugary gum from acacias provides tree rats with both food and a source of metabolic water.

Below, left: Black rhinos usually browse on a wide variety of woody plants and succulents, including some that are poisonous to other species.

Below: Giraffes are highly selective feeders, capable of selecting leaves and buds from among acacia hooks and spines.

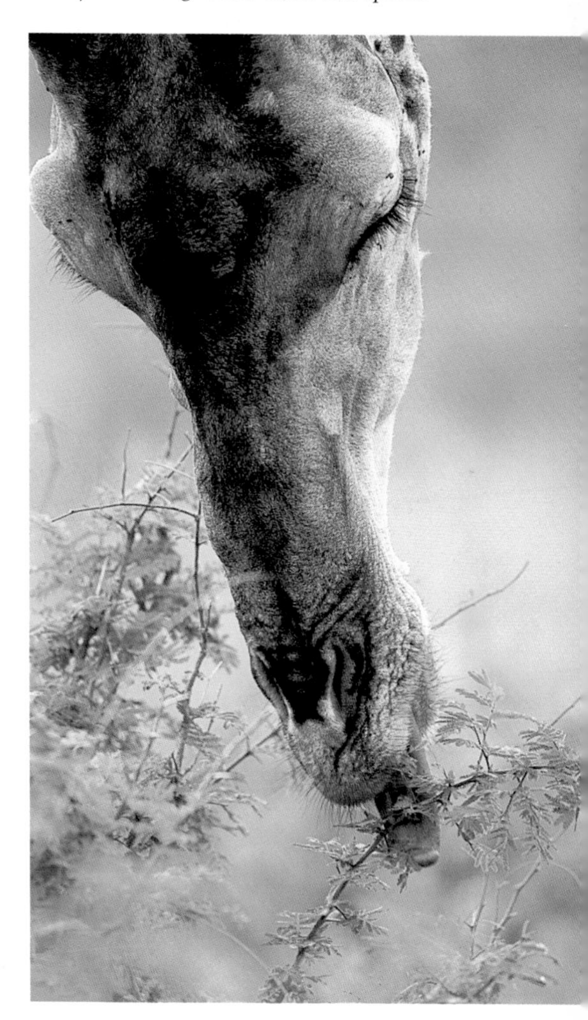

Top: A giraffe chews bones to supply its need for large quantities of calcium and phosphorus.

Above and opposite: To make up for their food's low content of essential minerals and sodium, most herbivores eat soil for the salts that it contains. This behaviour is illustrated here by black-faced impala (above), steenbok (opposite, top), and elephants (opposite, bottom).

breakdown. Instead, all herbivorous mammals depend on bacteria in their guts to convert cellulose into smaller molecules that can be absorbed and metabolised. The young of warthogs and zebra inoculate their guts with the bacteria that they need to help digest their food by eating the dung of adults. The co-adaptations of the bacteria and their mammal hosts are mostly hidden away inside the animals in the structure of their guts, but they surface in observable behaviour in two places: refection by rabbits and hares, and chewing the cud in ruminants.

A ruminant feeds fairly quickly, giving each mouthful only a cursory chewing before swallowing into the first of its stomach's four chambers. Then, either standing or lying down on its brisket, often in shade or thick cover for protection from heat and predators, it begins to chew the cud, regurgitating a mouthful of the coarser fraction of the rumen contents, chewing it with rhythmic sideways grinding of its lower jaw, and then swallowing it again for another cycle of bacterial processing. The repeated cycles of chewing and digestion reduce the food to microscopic fineness and convert the cellulose to small molecules that are easily absorbed. The dung of a ruminant can be distinguished at a glance from the coarser and more fibrous droppings of non-ruminants.

While the primary function of rumination is the digestion of cellulose, it is tied to an ability to recycle proteins and water that allows ruminants to penetrate into habitats that are too dry or where the forage is too poor for animals with less efficient digestions. Nonetheless, rumination is slow, and if the protein content of forage drops too low it is impossible for a ruminant to process food fast enough to obtain an adequate protein intake. Non-ruminants like zebras can survive on poorer quality food by passing larger quantities of it more quickly through their guts.

Rabbits, hares and vlei rats indulge in a behaviour known as refection – they eat faeces in order to obtain the nutrients that have been formed or liberated by bacterial action on the food during its first passage through the gut, but that have not been absorbed. Rabbits and hares feed at night, and during the day produce soft green faeces which they eat immediately. During the second passage through the gut the nutrients are absorbed, water is resorbed and the wastes are passed out as the familiar round, dry, rabbit pellets while the rabbit or hare is out grazing the following night. Vlei rats eat faeces, both their own and those of other vlei rats, from piles that have already been voided. Shrews, although they eat invertebrates, not plants, also use refection as a way of harvesting extra nutrients from their food.

There are important differences between grasses and broad-leaved plants in the nutrients and defensive chemicals that they contain, and these differences call for specialisations in herbivores' digestive systems. As a result, not many herbivores eat a mixture of grass and broad-leaved plants – those that do graze and browse, like impala, switch seasonally between the two rather than mixing them, turning to browse in the winter when the grass has been grazed down or becomes too poor in nutrients. An eland or grey rhebok feeding in long grass may look as if it is grazing, but feeding in grassland is not the same as feeding on grass, and careful observation will show that it is selecting broad-leaved plants from among the grass.

Because the bacteria in a herbivore's guts that help it to digest the cellulose in its food are adapted to its diet, it is essential that when animals are introduced to a new area, where the food plants are different to those they have been feeding on, they are gradually acclimatised to their new diet before being released and left to their own devices.

Browsers such as kudu or giraffes spend less than about five minutes feeding on each tree before they move to the next one, and they move on when there are still plenty of leaves available where they are. There is a story that they are forced to move on by the rapid build-up of distasteful and toxic tannins in the leaves of the tree, but plants simply cannot respond that fast. A far more likely explanation is that it is more difficult for a predator to stalk an erratically moving target than one that stays in one place, steadily munching. The slow pace of plants' physiological responses, and experiments that consistently fail to reproduce the effect, also give the lie to the story that a tree that is being browsed warns its neighbours to mount their own chemical defences.

Above: This hippo cow's nudging and nibbling of a dead baby hippo may reflect a taste for meat. This is an interesting behavioural atavism because hippos' teeth are completely unsuitable for chewing flesh – even when they graze they use the edges of their lips to crop the grass; their canine and incisor teeth are specialised as weapons for fighting, not for feeding.

Many herbivores eat soil – a behaviour with the somewhat grand title of geophagy – in order to obtain the trace elements that are lacking in their diet and the sodium that they need to keep their rumens functioning. For the same reason, most antelope prefer to drink slightly brackish water. Giraffes have a huge demand for calcium and phosphorus to build their skeletons, and where the soils and therefore the vegetation are deficient in these elements they obtain them by chewing bones. Doing so exposes them to the risk of contracting anthrax or botulism, whose resistant spores can persist for years even in dry bones.

Some herbivores take to eating animals if vegetable food is scarce; hungry ground squirrels eat termites, and hungry multimammate mice eat each other. Hippos, although they have the body and digestion of grazers, seem to be haunted by the omnivorous ghost of a pig-like ancestor. They show an unexpected interest in carcasses floating in the water where they spend the day, and observations in Hwange National Park during the drought of 1995 confirm that they eat some of the meat.

Because greenstuff is not particularly nutritious, the large grazers and browsers are committed to having to feed for long periods in order to ensure that they get an adequate intake of nutrients. White rhinos graze for 12 hours a day, elephants feed for 16 hours and zebras for up to 19 hours. Kudu spend more than half their time foraging, and if the time that they spend ruminating is added to that, feeding behaviours take up more than three-quarters of their lives. These hours of patient reaping are in striking contrast to a large carnivore's lifestyle of 2–4 hours of hunting, a couple of minutes to capture and kill, less than an hour to feed, and then another 18–20 hours of indolence.

Previous page: *A lioness needs only a few hours of activity to secure her daily food requirements.*

Below: A bat-eared fox detects its food by sound and it can locate beetle grubs buried as deep as 30 cm under the surface.

PREDATORS

With only a few exceptions, predators are faced with the problems of detecting, pursuing, capturing and subduing their prey.

A golden mole detects its prey of invertebrates or – for Grant's golden mole – small lizards by sound, touch and probably smell. Elephant shrews locate insects by sight or by sniffing into crevices with their long, incessantly mobile snouts. Scavengers detect food from the sound of predators on a kill, by watching for vultures swooping down, and by catching the scent of a carcass on the wind. Spotted hyaenas in the Kalahari have been seen to turn upwind towards a carcass lying 4 km away.

Aardvarks feed on termites, but instead of waiting for them to emerge onto the soil surface an aardvark digs them out. It detects underground workings by smell, and when it scents a likely spot, makes a shallow exploratory scrape in the soil. If the prospects are good, it digs deeper until it uncovers the tunnels or nest, and then licks up the termites by pushing its long cylindrical tongue 20–30 cm into the termites' tunnels. With above-ground nests the aardvark's approach is more direct; it rips a hole in the wall and digs towards the heart of the nest, feeding as it goes. The mud and saliva mixture that the termites use to build with sets hard enough to turn a pick-axe: it is testimony to an aardvark's strength that it can dig a hole big enough for its bulky body to fit into.

All predatory carnivores use sound to help locate their prey, and in two species the ability to pinpoint exactly where a sound is coming from is so finely honed that they can capture prey that they cannot see. Serval are lightly built, long-legged cats whose favoured habitats have dense, long grass in which they hunt for vlei rats and other rodents. In dense, matted grass a serval may not always be able to see its prey, but its large ears can localise the sound of a rat's movements so accurately that the serval is able to pounce with deadly accuracy in a high arc that may carry it four metres over the intervening grass.

Bat-eared foxes are the least carnivorous of the African canids – most of their diet is supplied by termites, grasshoppers and dung beetle larvae. The first two are detected by sight and sound, and if a grasshopper remains motionless it has a good chance of escaping notice. Dung beetle larvae feed in dung balls buried underground, and a foraging bat-eared fox detects them by the sound of their gnawing at the inside of their dung ball nurseries. The fox walks slowly with head tilted forward, and when it catches the sound of a grub it lowers its head until its ears are almost touching the ground and moves its head and ears until it has precisely located the source of the sound. Then the fox begins to dig, and its pinpointing of where the grub lies is so accurate that it can dig straight down to it without wasting time and energy on a hole that is any bigger than is strictly necessary.

Coalitions of male cheetahs are the only cats other than lions that regularly hunt together.

An aardwolf forages by listening for the sounds that betray the activities of an above-ground band of harvester termites. When the aardwolf locates a band it has a unique and specialised way of capturing its members, licking them up with strokes of its broad and sticky tongue that are too fast for the human eye to follow. It has a need for speed: the faster it can lick, the more worker termites it can capture before the numbers of foul-tasting soldiers increase too much. With an intake of a quarter of a million grass-eating termites a night, and 100 million of them a year, an aardwolf is one of the best allies a stock farmer could have.

Cape clawless otters are hand-focused – they find and catch prey with their dextrous forepaws and when on land they carry food by clamping it against the chest with one forepaw. Spotted-necked otters are mouth-focused, they catch fish with their teeth, and carry food in their mouths.

Considering the violent turmoil at a kill, some of the techniques that predators use to secure their prey are quite astonishingly clinical. In some wild dog packs, one dog will immobilise a large grazer by gripping it by the nose, which has the same paralysing effect as a twitch on a horse. In the southern Kalahari spotted hyaenas snap at their victims' eyes, leaving clear signs on the skull.

Cheetahs' hunting style has one requirement that overrides all others – the ability to accelerate from a standing start to a top speed of 100 km an hour. Natural selection has left them long-legged and lightly built; they stand 20 cm taller than a leopard but weigh 5–10 kg less. Their heads are small and their canine teeth are short – wide air passages have taken over the spaces where the roots of the teeth would be. Their tails are long and flattened, acting as both rudder and counterweight as the cheetah matches its target's evasive turns.

Although its extended high-speed chase is distinctly un-feline, a cheetah is just like other cats in that it stalks as close as it can to its prey before launching an attack. When stalking over open ground it slinks forward whenever its selected target drops its head to feed, and freezes into immobility as the prey looks around for danger.

Attacks are most successful if they can be launched from less than 30 metres, while the prey is still unaware of the cheetah's approach. It can then be charging at full speed even as its target becomes aware of the danger. But if the prey breaks and runs before the stalk is complete the cheetah may still attack from as much as 100 metres away. Between a third and half of all chases are successful, but three-fifths of stalks are aborted because the target sees the cheetah and runs off.

Opposite: Cheetahs are specialist sprinters; the only carnivores that outpace their prey in acceleration and speed.

A cheetah's explosive acceleration, high speed and manoeuvrability demand better traction than a padded foot alone can provide, and so its claws are permanently extended. Blunted by contact with the ground, they are no longer used to grapple the prey; that function is now performed exclusively by the dew claws, which are compensatingly larger, a 25 mm hook just below the ankle joint on each foreleg. Drawing up behind its prey at 100 km an hour, a cheetah breaks stride and slaps forward and sideways with one or both forelegs, hooks a dew claw into the prey's rump or flank and jerks backwards to spin the prey off balance and into a tumble that ends with the cheetah's teeth clamped in the prey's windpipe. Its exceptionally wide nasal passages allow it to maintain a throttling grip while sucking copious breaths of air into its own lungs, both to balance its oxygen debt and to shed the heat generated by the explosive effort of the chase.

Above: A cheetah nearly always kills with a suffocating throat bite – its canine teeth are too short for the bite to the back of the neck that other cats sometimes employ.

As soon as possible after killing, a cheetah drags its prey into cover to escape the notice of vultures, which are not only competitors in their own right but also attract the attention of lions and spotted hyaenas, which frequently steal cheetahs' kills. With its lightweight skull and teeth a cheetah has difficulty biting through the skin of even small antelope, so it usually starts feeding at the groin where the skin is thinnest. It does not eat the skin or intestines of its prey and can chew up only small bones with its weak teeth.

Because it is so perilously dependent on perfect physical fitness, a cheetah dare not risk injury in confrontations with other predators, and it will abandon its kills or even its cubs when challenged by lions, leopards or hyaenas. Even warthogs and wild dogs have been seen to chase cheetahs off a kill.

Wild dogs are considered the most purely carnivorous of the canids. They are specialised hunters of medium-sized antelope such as springbok, impala and young blue wildebeest, but they also take animals as small as mice and as large as kudu and zebra. They hunt by sight, usually in the cool of early morning or evening, and alter their hunting behaviour to suit the prey and the habitat. They use what cover is available to sneak up close to prey, but will also break into a chase from an unconcealed approach. When after large prey the pack singles out an individual target, but with smaller species like impala that scatter when attacked the pack splits up and may make multiple kills.

Above: Wild dog pups beg an adult for a meal of regurgitated meat.

Bottom: When food is abundant spotted hyaenas gorge themselves into near immobility. A soak in shallow water helps to dissipate the heat generated by digestion.

Opposite: By far the most arboreal of the African cats, leopards use trees as refuges from other carnivores and as larders for their kills.

Wild dogs can reach speeds of 65 to 70 km an hour, too slow to catch medium-sized antelope in a short chase – for that the cheetah's acceleration to 100 km an hour is required – and they depend on their stamina in long pursuits to wear down the prey. If they have to, the dogs can keep up a speed of 60 km/h for 4–5 km. If the prey circles, as territory holders often do as they attempt to stay on familiar ground, the trailing members of the pack gain on it by cutting the corner, but it is usually one of the alpha pair that leads the chase from start to finish, and is first in at the kill. The dog that catches up with the prey grabs it and pulls it down if it is small enough, or runs alongside and slashes at its rump or shoulder to slow it down while reinforcements arrive. Although wild dogs do not have a specialised killing bite like the cats, their method of dispatching the prey is just as effective and just as quick; they bite open its belly and tear out its heart and lungs.

When a kill has been made the juveniles rather than the adults are the ones who feed first. Carcasses are consumed quickly, with none of the snarling and bickering that goes on at lion or spotted hyaena kills. This reduces the risk that kills will be stolen by larger predators or vultures. A pack of wild dogs is usually able to keep hyaenas at bay because some of the dogs stand guard while the others feed. The pivotal role of the pack in the survival of wild dogs comes to the fore after the hunt, with the division of the spoils. A dog that has remained behind at the den to guard the pups, lost the trail of the hunt, or stood guard at the kill begs for food from other pack members by whining and nudging and nibbling their lips and licking their faces. In response, dogs that have fed regurgitate lumps of meat, sometimes directly into the supplicant's mouth. Weaning puppies are fed in the same way by the adult dogs, even by those who had to beg for the food that they give to the youngsters. So strong is the urge to share food that a single chunk of meat might pass from one dog to another through as many as four or five stomachs. Sick and injured wild dogs also receive a dole of meat as long as they remain with the pack.

The larger carnivores, from cheetah upwards, live by boom and bust. If they kill or scavenge something large they will gorge themselves, but may then have to go for days before they get another meal. A lion can swallow 35 kg at a sitting, and a lioness 22 kg, which is 15 per cent of their body weight, and enough to satisfy their average daily intake for a week. A spotted hyaena can bolt down about 20 kg, 30 per cent of its body weight and six times its average daily intake, in less than 15 minutes, and then go for five days without food.

Where there are no scavengers large enough to pose a danger, a leopard will leave its kills on the ground, possibly covered with dirt or branches. In areas where lions and hyaenas occur, leopards protect their kills from scavengers by taking them up trees. With its store of meat well protected from competitors a leopard is able to feed on a carcass for three or four days, by which time its tolerance for rotten meat will have been sorely tested. Only hyaenas are more willing than leopards to feed on decaying flesh.

Jackals and hyaenas of both species cache stores of food whenever it is abundant enough to satisfy more than their immediate hunger. A brown hyaena that found a nest of 26 ostrich eggs ate seven of them on the spot and then carried the rest away one by one to hide them in grass clumps and under bushes up to a kilometre away. Spotted hyaenas often cache meat in shallow water, presumably to keep it safe from vultures and to hide its scent from other scavengers.

Opposite page: With its grey duiker kill safely stashed, a leopard has enough food for the next two or three days. Not only small antelope carcasses are stored in this way, there have been reports of young giraffe weighing more than 100 kg being hauled more than five metres into trees.

BATS

The vast swarms of insects that fly at night offer a feast of high-quality, nutritious food to any animal that can locate and capture a small flying target in the dark. Insect-eating microchiropteran bats have a virtual monopoly over exploitation of nocturnal insects because they can echolocate – the bats produce pulses of very high frequency sound and listen for the echoes that bounce back from objects in their paths. The further away an object is, the further the outbound pulses and their echoes have to travel, and the bat judges distance from the time lag between call and echo.

Only objects that are large in relation to the wavelength of a sound can produce an echo, and for sound to echo from something as small as an insect it has to have a very short wavelength, which means a very high frequency. Bat echolocation ultrasound has frequencies as high as 210 000 Hertz (a Hertz, abbreviated Hz, is one vibration per second), which is 10 times higher than the upper frequency limit for human hearing. The higher the frequency of a sound, the shorter the distance that it carries, and a 100 000 Hz ultrasound pulse dwindles to one fifty thousandth of its original intensity within only five cm. By the time an echo gets back to the bat it is more than a million times less intense than the sound pulse that the bat sent out.

For the echoes to be loud enough to hear, the bats have to shriek their echolocation pulses at intensities that are difficult to imagine. The sound level 5–10 cm in front of an echolocating bat is often around 170 decibels (dB) and can be as high as 300 dB even 15 cm away. For comparison, the sound pressure from a gunshot is

110–140 dB, and a lion's roar reaches a paltry 114 dB. Producing these blasts of sound demands such an effort that bats synchronise their calling with their wing beats. Bats that fly close to the ground or among dense vegetation do not need far-carrying calls and so they whisper at less than 54 dB; the loudest calls come from bats that fly high and fast. So that the whispers of the returning echoes are not drowned by the shrieks of the outgoing pulses, a bat alternates between calling and listening.

The shapes of echolocation pulses are adapted to the conditions under which different sorts of bats hunt. Pulses with a constant sound frequency help to pick out an insect's regular wing beats against a noisy background of wind-stirred foliage. Commonly the sound frequency during a pulse is swept downwards, which may help the bats to identify different targets by the way that they reflect different frequencies. Some bats use pulses with both fixed frequency and swept components.

Opposite, above: A baby suricate waits for its caretaker to catch its next meal.

Opposite, middle: Giraffes are very vulnerable when they drink and always look around carefully and make sure they have a firm footing before bending. The brain is protected by a special system of blood vessels from the sudden changes in blood pressure that come from raising and lowering the head.

Opposite, bottom: Careful management of water supplies is crucial to the conservation of roan antelope. If there are too many waterholes the roans' grazing is destroyed by zebras and blue wildebeest.

The pulses are very short – a bat flying high and fast gives a loud pulse about once a second, probably to check its height above the ground. When cruising where there are obstacles a bat orientates itself with pulses between 0,2 and 100 milliseconds long at a rate of about 10 per second. If it hears something interesting its sound pulse rate rises to 20–50 per second, and climbs to 200 per second as it closes on its prey. As the pulses follow each other more and more quickly the echo from one is coming in while the bat is giving the next one, and so the pulses are shortened to leave more time between them for listening.

The noseleaves that horseshoe, slit-faced, leaf-nosed and trident bats carry on their faces serve as sound reflectors, beaming the echolocation pulses forward as the ultrasound equivalents of a searchlight. This serves both to increase the intensity of the sound and to keep the outgoing pulses away from the bat's ears. The tragus, a small flap of membrane at the ear opening, may also help to deflect the direct sound of the bat's own calls. Most microchiropterans have large and exceedingly mobile external ears to capture returning echoes. By exploiting the directionality of its ears a bat can distinguish the echoes of its own calls from background noise that is 2 000 times louder. Some of the bats with exceptionally large ears may be able to locate prey by the sounds that it makes itself, rather than by having to use echolocation.

Once an insect has been located and pursued it may be scooped up in the tail membranes, fielded with a wing or snapped out of the air. Insects – such as arctiid moths – that are foul-tasting use an ultrasound equivalent of warning coloration to repel hunting bats. When a moth picks up a bat's echolocation pulses it emits ultrasonic clicks of its own, which warn the bat not to pursue matters further.

Some bats eat on the wing, others fly to a perch where they can hang up while they manipulate the prey with the claws on their wings. Large and common slit-faced bats and Geoffroy's horseshoe bats litter the ground beneath their feeding perches with the inedible fragments of their insect prey.

Marine mammals, like toothed whales and dolphins, also make use of echolocation to locate prey, and to find their way in dark or muddy water. Pygmy sperm whales and sperm whales can produce sounds that are so loud that they stun the squid that they feed on. Because the wavelengths of sound are longer in water than in air, whales and dolphins have to use frequencies even higher than those used by bats – sometimes up to 300 000 Hz – but ultrasound carries much more strongly through water than through air and so it works over longer ranges.

TO EACH ITS OWN

Co-operation in feeding is remarkably rare, even among the members of social groups. In a herd of antelope or a troop of baboons or monkeys, each animal forages for itself. A subordinate baboon that finds a rich food source – perhaps a large root or a pile of dung full of beetles – hurriedly stuffs its cheek pouches and retreats to a safe distance from the rest of the troop to avoid being robbed of its spoils. Even suricates and dwarf mongooses, which are so conspicuously social in other respects, co-operate in catching and dispatching prey only when it is exceptionally large, such as a big lizard, or when a caretaker (*see* Chapter 5) brings food to its charge. In contrast, a banded mongoose that finds a rich patch of food such as a dung pile full of beetles and grubs will give excited churrs and twitters that attract its companions.

When an eland or kudu bull pulls down a branch with its horns, others in its herd may feed on the leaves, but like cows and calves eating the camel-thorn pods knocked down by bull elephants, this is exploitation rather than co-operation. In fact, collaborative foraging is more or less restricted to the large carnivores that hunt in groups – lions, spotted hyaenas, wild dogs and cheetahs. Nonetheless, in a Mana Pools National Park rest camp a ratel was seen to squeeze through the narrow opening of a rubbish bin and then throw food scraps out to its partner, and in an astonishing observation in the Kalahari, photographed by Peter Pickford, one ratel fed another that had a broken back.

WATER

Water is the universal solvent in which all the processes of life take place. Animal life's invasion of dry land depended on the evolution of skin that was waterproof enough to stop water evaporating away into the air. The skin of nearly all mammals is waterproofed by waxy and oily secretions produced by apocrine sweat glands, but mammals still have to take in water to balance what they use and lose. All terrestrial vertebrates lose water from their lungs by evaporation during breathing, and in the essential elimination of wastes as urine or faeces, but mammals have two uses for water that are unique: their bodies are cooled by the evaporation of watery eccrine sweat produced by tiny glands in their skins, and mothers' milk is 40–95 per cent water.

Most of southern Africa is semi-arid, and water is a critical resource for a wide range of mammals.

In southern Africa there are very few habitats that provide an unlimited supply of water, and southern African mammals have a variety of adaptations that allow them to balance their water requirements against their water intakes.

While they are cropping the lush grass of spring and early summer, most grazers do not need to drink, or can go for three or four days between visits to water, but as the season progresses and the grass grows coarser, then dies back and dries through the winter, the importance of regular access to drinking water grows. The distribution of most mammal species is constrained by their need for drinking water. Reedbuck are found only in wet habitats where they can drink three or four times a day. Sitatunga, red lechwe and puku are even more tightly tied to water, and their only stronghold in the subregion is the seasonally flooded habitats of the Okavango-Linyanti-Chobe system in the far northern parts of Botswana. For sitatunga and red lechwe, water is required not only for drinking and to provide green grazing, but also as a refuge from potential predators.

Wild melons are an essential source of water for ground squirrels and dozens of other species in the Kalahari and Namib.

Except during the early summer when lush grazing is available, those ubiquitous grazers Burchell's zebras, impalas and blue wildebeest are restricted in their movements by their need to drink daily. Under natural conditions this excludes them from large areas that are utilised by other species that are less water dependent, but cannot compete effectively for food. South Africa's Kruger National Park is only now being managed away from the ecological distortions that resulted from the well-meaning, but with hindsight sorely misguided, sinking of hundreds of wind-pumped boreholes during the 1960s. When there is permanent water available all over the Park there is nowhere that zebra and blue wildebeest cannot feed without getting beyond the 12 km radius from water that allows them to drink every day. Their grazing down of tall grasslands drove the Park's roan antelope to within fewer than 56 animals of the precipice of extinction, because roan cannot feed effectively on grass that is less than 10 cm tall and they need habitats with grass 0,5–1,5 m tall for their calves to hide in for their first six weeks if they are not to fall easy prey to predators.

Water comes in many guises besides its liquid form; throughout the southern Kalahari tsama melons provide the main source of water through the dry winter months. Because their food has a high water content, carnivores and insectivores can go for long periods without drinking, but they will all drink if water is available. Kalahari cheetahs drink the blood and urine of their prey and eat tsama melons, and they can travel 80 km between drinks of water. Yellow mongooses lick the early morning dew from plants, and also eat succulents. Although baboons typically have to drink every day, baboons in the Kuiseb River Canyon in the Namib Desert regularly go without water for up to 11 days at a stretch, and once survived for 26 days without drinking by resting for hours in the shade and selecting food with a high water content. Brants' whistling rats can survive under desert conditions by eating fresh vegetation, even though their physiology is not particularly well adapted to dry conditions. Tree rats in the southern Kalahari obtain water by eating acacia gum, and to reduce their evaporation losses they increase the relative humidity in their nests by urinating over them. Jackals on the Namibian coast eat melons and lick settled fog. Samango monkeys lick rain water from leaves and drink from tree holes.

As long as they have green forage a surprisingly wide range of the larger herbivores are independent of water. Giraffes, dassies, klipspringers, suni, Damara dik-diks, oribi, steenbok, eland, nyala and grey rhebok can live without drinking as long as they have greenery to eat. Steenbok can even survive by digging for succulent roots. All of them, with the surprising exception of oribi, will drink if water is available.

Gemsbok and springbok are true desert antelope, which can survive indefinitely without drinking. They eat succulent underground stems and roots and wild melons, and conserve water by producing very dry faeces and concentrated urine. Rather than use water for cooling by sweating they use behavioural methods of keeping cool such as standing in shade during the day and grazing at night when the water content of the vegetation is higher – even dead grass that is baked to biscuit dryness during the heat of the day can pick up 30 per cent of its weight in water as temperatures drop and relative humidity rises after dark. A gemsbok or eland allows its body temperature to rise during the day, and then it cools down again at night, often moving up to the crest of a dune to catch the evening breeze. A gemsbok's maximum body temperature of 45 °C is actually high enough to cause brain damage, but its brain is protected by its blood supply being cooled by heat exchange with cooler blood flowing back from the nasal membranes.

Opposite top: A diet of fresh greenstuff allows Brants' whistling rat to survive without drinking water.

Opposite, bottom: Steenbok can survive indefinitely without drinking by choosing the freshest herbage, eating wild melons and, like this one, digging for roots.

Above: Carnivores like this leopard can obtain a large portion of their water requirements from the body fluids of their prey, but they will all drink if given the opportunity.

Highly mobile antelope species such as blue wildebeest and red hartebeest, for example, are able to migrate very long distances when rain fails in one area and falls in another. Their long forelegs and high shoulders allow them to move at a canter that is faster than normal walking and also uses less energy than trotting or galloping. These long-distance movements were once a feature of the drier western two-thirds of the subregion but fences now block the migration routes across the whole of the subcontinent.

Lack of water and shortage of food are inexorably linked. If the rains fail, grass does not grow, and to provide an artificial water source does not always solve the problem; it simply means that animals first destroy the vegetation around water points, and then die from starvation instead of thirst. This happened in the former Kalahari Gemsbok National Park in 1985, when eland, red hartebeest and blue

After allowing their body temperatures to rise during the day, gemsbok climb to the crests of dunes to take advantage of the cooling evening breeze.

wildebeest trekked southwestwards out of the Kalahari region and starved in their thousands around the artificial water points along the Nossob and Auob river beds.

When no surface water is available for drinking, gemsbok, zebras and black rhinos dig for it in dry river beds, but the greatest well diggers are elephants, and they have a remarkable ability to locate where the underlying water comes closest to the surface. Most likely they can smell moist soil, because they dig up and wreck any water pipe that has even a tiny leak. They dig two sorts of holes – large basins that they scrape out with their feet and that can be used by other animals after the elephants' requirements of 50–200 litres a day each have been satisfied, and deep, narrow holes, often next to a clump of grass, that they dig by scooping sand out with their trunks. Only elephants can reach the water that seeps into the bottom of these wells.

CHAPTER FOUR

4

STRATEGIES FOR SURVIVAL

challenges and adaptations

ANIMALS CAN DIE FROM HUNGER, thirst or disease, in fights, by accident, or in the jaws of predators. In nature very few of them are spared to die of old age. Survival is an immediate imperative but not an end in itself; rather it is a means to another end, the reproduction that passes genes on to subsequent generations.

ELUDING THE ENEMY

Of all the threats to survival, attacks by predators are the most immediate and the most dramatic. The first step in not becoming a predator's next meal is avoidance, and a large majority of mammals spend at least their resting periods in refuges that are inaccessible to predators. Insectivorous bats and Egyptian fruit bats hang on the walls and ceilings of caves, or tuck themselves into rock crevices and tree holes. Most rodents sleep in burrows, emerging to feed; giant rats and pouched mice lay in larders of food as insurance against times when a predator's activities make foraging too dangerous. Golden moles and mole-rats even forage underground, and can live their whole lives without ever risking coming to the surface.

The first line of predator evasion for animals that have no secure refuges – which includes hares, rock rabbits, zebras, all the antelope, and even the larger carnivores themselves – is concealment; the art of not being detected. Keeping to heavy cover and wearing camouflage are elements of concealment, but its essence is to remain completely still. In the broken shade of a forest floor or savanna woodland the plain grey of an elephant or duiker can blend as effectively as the jigsaw blotches of a giraffe, as long at their wearers keep from moving. Hiders sneak away before being detected, or explode from cover when discovery is imminent, buying a fraction of a second's advantage before pursuit begins. For animals like greater cane rats, which live among very dense grasses and reeds, a silent retreat or flight is almost impossible. The cane rats incorporate this into their escape strategy, crashing away at high

Previous pages: Jackals are just one of the hundreds of species of mammals that use burrows as refuges from predators, and for shelter from the weather.

Below and below, right: The immobility that allows potential prey to blend with their surroundings is critical to avoiding detection by predators.

Large trees, cliffs and caves, which are relatively secure against predators, are used by baboons as night-time refuges.

speed and then coming suddenly to a silent stop that makes their movements very difficult to follow. Animals that live out in the open also have little chance of making a discreet departure – if they have detected the predator it is almost certain to have detected them.

No matter how pressing the threat, potential victims have to strike a balance between safety and getting on with the other aspects of their lives. A mouse down its burrow, a bat hanging in a cave, or a baboon roosting on a cliff ledge are certainly safe from predators, but only at the cost of not feeding, drinking or looking for mates. Antelope and zebra simply cannot afford to waste feeding time and burn up energy by fleeing at every small alarm – and there is always the danger, especially if lions are the hunters, that flight from one predator will lead to an ambush by another. Even maintaining a vigil against approaching danger cuts into the time that is needed for other activities; every time that a grazer lifts its head to check the surroundings it loses a couple of bites of forage, and with feeding already taking up more than half their day herbivores have no choice but to risk predation in order to feed. The risk may be substantial: cheetahs are expert at stalking where there is no cover by sneaking forward when the target lowers its head to graze and freezing into invisible immobility as it raises its head to scan for danger.

Zebras' stripes may distort perspective, and make it difficult to pick an individual from a herd.

ZEBRA STRIPES

Avoiding predation is one of the many functions that have been postulated for zebras' stripes, but some of the models for how the stripes fulfil that function clearly do not fit the behaviour of zebras and the predatory pressures that they suffer. Others are more plausible. One suggestion – that the stripes provide camouflage in long grass, or even in the dappled shade beneath bushes and trees – falls foul of zebras' preference for areas of short grass, and although the stripes do break up their outlines, they are still easy to see against any background. And zebras do not behave as if they are camouflaged: they are active and noisy, and when they detect a predator they give loud alarm calls. In complete contrast, an animal like a kudu or bushbuck, which does rely on camouflage to avoid predators, is stealthy and quiet, and favours dense, not open, habitat. At the approach of danger it freezes into immobility, and only flees at the last moment.

The pattern of a zebra's stripes – narrow on its head, neck and shoulders and widening towards its rump, and cutting the outline of the body rather than following it, may affect a predator's judgement of the zebra's size, how far away it is, or exactly which direction it is facing or moving in. The slimming effect of vertical stripes on a dress is well known to all fashion designers, while the horizontal stripes on a rugby jersey make its wearer look more bulky and formidable, and it is possible that the same effects could be produced by zebra stripes. Precise judgement of distance, direction, speed and size is critical for a pursuing predator to capture and bring down its prey, but if a zebra's stripe pattern does affect any of these it is not in any very dramatic fashion. Zebras fall prey to a wide range of predators with very different hunting methods: lions and leopards that stalk and rush, spotted hyaenas that chase over long distances at night, wild dogs that do the same in daylight, and cheetahs that chase at high speed over short distances. Similar arguments apply to the possibility that the chaos of stripes in a herd of zebra make it difficult for a predator to pick out and focus on an individual target.

Herbivores of other species are preyed on by the same predators that prey on zebras, and yet none of them have anything even close to a zebra's stripes. Zebra stripes are unique: although members of other species – kudu, nyala and bushbuck, for example – also have stripes, they are nowhere near as bold as the high-contrast black and white of a zebra's pattern, and they cover only a part of the body. The stripes in these species are almost certainly for camouflage, their wearers live in dense cover, and have the stealthy habits that would be expected of animals that strive to avoid detection. The large herbivores that share the zebras' open habitats are either plain-coloured, like blue and black wildebeest, red lechwe and red hartebeest, or strikingly marked in ways that exaggerate their body outlines, like gemsbok and springbok. So, assuming that the stripes are an anti-predator device, is there anything unique about zebras' anti-predator behaviour that connects with their unique pattern? Zebras live in breeding groups or bachelor herds (*see* Chapter 5).

Zebras need to stay closely bunched when fleeing predators; their high-contrast black and white stripes make it easy for group members to see one another so that they can manœuvre as a unit.

Each breeding group consists of a stallion with his harem of mature mares and their foals. When predators attack, a zebra group bunches tightly and flees with the stallion in the rearguard and the mares with the youngest foals in the position of greatest safety at the front of the group. The stallion manoeuvres on the flanks and rear of his group, kicking and biting at any pursuer careless or reckless enough to come within range of his hooves and teeth. This group formation is a very effective defence and predators depend for success on being able to cut a single animal out of the bunch. As long as it can stay in the group an individual's chances of being captured and killed are lower. Fleeing in a tight bunch with a defensive rearguard is as unique to zebras as their stripes are – the connection between the two is the circumstances under which the bunching strategy has to be brought into play. To be able to stay tightly bunched the zebras need to be able to see, and be seen by, one another, but most predatory attacks happen after dark, and visibility is reduced still further by the clouds of dust thrown up by fleeing hooves. Under these conditions a pattern of bold black and white stripes offers its wearer the best chance of being seen so that its companions can more easily stay close to it.

Below: Before the end of the first week young springbok sprint away if threatened, but it takes a month before they can match the speed of an adult, which would have been this doomed lamb's only chance of survival.

Opposite: Explosive acceleration enables medium-sized antelope to sprint away from pursuers, but they are vulnerable in long chases.

FLIGHT

Among carnivores, by far the most common hunting tactic is the stalk, rush and pounce used by cats, mongooses and genets. A cheetah's sprinting run is an extended rush. Extended chases that are tests of stamina and endurance for both predator and prey are used by only two species of African carnivore, wild dogs and spotted hyaenas. Whether a predator's attack ends with the prey's capture depends on more than just raw speed, even when the pursuer is the world's fastest land animal, a cheetah. The target's ability to jink and dodge can shake its pursuer loose; the agility to match its target's evasive manœuvres can fill a predator's belly. A dodge only works if it comes at the very last moment, as the predator closes from behind and commits itself, so that it overshoots as its target cuts away. Any earlier and the pursuer actually benefits, simply cutting the corner and leaving the dodger worse off than it was before. This is why hares run in front of vehicles, sometimes for hundreds of metres, before dashing to one side, seemingly – and sometimes actually – right under a front wheel.

Members of several species – among them reedbuck, blesbok, Damara dik-dik, oribi and even scrub hares – sometimes flee from predators with an exaggerated, bouncy gait that is slower than the flat-out sprint that would be expected under the circumstances. These exaggerated gaits are known collectively as stotting. Their details vary with the species: grey rhebok and scrub hares run with a rocking horse action, Damara dik-dik bounce rhythmically along, a reedbuck kicks backwards at

If one of a klipspringer pair detects approaching danger it alerts its mate with a loud whistle while standing prominently on a boulder.

the crest of each leap. In springbok, stotting reaches such a peak of spectacular extravagance that it has been given a special name – while other species stot, springbok pronk. Pronking involves vertical leaps with the legs held stiffly straight and the feet close together, head down, and back arched. Variations on the theme include the classic pronk with the head held high, a high-stepping gait similar to a horse's triple, exaggerated long paces or spectacular leaps in which the springbok takes off, paces in the air and comes down running without even breaking stride.

Springbok pronk in a variety of circumstances: when a herd is trekking to food or water, when there is some mild disturbance, just before a change in the weather, and sometimes for no apparent reason at all. The most puzzling thing about pronking is that it is also a response to a potential or actual threat of predation, or even to being pursued, when it would appear that the most appropriate course of action would be to flee immediately at top speed.

In all species that stot the action is accompanied by the flashing of a white rump patch, or the white underside of the tail. In the full pronking display a springbok flares the fan of long white hair that usually lies concealed in a fold of skin along its back. The skin at the root of the fold is glandular and it produces a sticky secretion with a sweet, heavy smell. The visual display of pronking is thus combined with an olfactory signal as the scent of the secretion is released. The sharp pop that coincides with the backward kick that a stotting reedbuck gives at the crest of a leap might be due to scent glands in its groin opening and closing, and adding an olfactory signal to the visual and audible displays of the stotting action.

Stotting and pronking pose something of a puzzle – why do animals that are under threat from a predator not simply flee at top speed instead of stotting? Some of the answers to that question have come from work in East Africa on Thomson's gazelles, the springbok's ecological equivalent.

In gregarious antelope, stotting certainly does act as an alarm signal to the rest of the herd, and in the resulting confusion of fleeing bodies the stotter has a good chance of escaping the attention of a predator. If threatened by a predator a young Thomson's gazelle will stot to attract its mother's attention and to get her to come to the rescue. But scrub hares are solitary, and have not been seen to protect their young and so their stotting cannot be an alarm signal to other hares, nor is it likely to be a plea for help.

Stotting might trick a predator into attacking too soon, while the prey is too far away to be overtaken in the first rush, and it has even been suggested that it gives the stotter a wider view of its surroundings as it reaches the crest of each leap, though this explanation cannot fit rocking horse gaits which hardly lift the stotter's head at all, or springbok, which pronk with their heads down.

When the evidence is put together it suggests that stotting and pronking are signals to the predator that it has been detected, and that its intended victim is agile, fit and fast, and not an easy target.

For a stalk and rush predator such as a cat, which cannot sustain a high-speed chase, being able to launch an attack before its target realises that there is any danger is so crucial to success that a predator whose presence and location is known is almost harmless. And so, antelope will actually follow lions around in order to keep them in sight. Klipspringers live in pairs, and if either member of a pair detects a predator it gives an alarm whistle while standing prominently on a boulder or other vantage point. The alarm whistles have two functions: a whistle from one member of the pair alerts its partner to danger, but also informs the predator that it has been detected, and that its chances of making a kill have been lost along with the element of surprise. Having been detected, the predator is likely to leave the area to hunt for less alert prey elsewhere. Even knowing that a sentinel is on guard may be enough to persuade a predator to move on, and tsessebe bulls keep watch for both rivals and predators while they stand prominently on a termite heap or other vantage point in their territory, with their harem of cows grazing nearby.

Brants' whistling rats are named for the high-pitched whistle that they give as they disappear into their burrows at the approach of danger in the shape of a predatory mammal or bird. If the threat comes from a snake, which can easily crawl into a rat's warren, the rat remains above ground and whistles repeatedly, which brings its neighbours out of their holes. The single whistle for an above-ground hunter and the repeated whistles for a snake are a simpler

A pronking springbok flares a crest of long, white hair, releasing a sweet heavy smell from secretion at the roots. While pronking, a springbok can sometimes clear three metres, four times its shoulder height.

version of the predator-specific alarm calls used by vervet monkeys, suricates and dwarf mongooses. Why the rats whistle at all is something of a puzzle, because they live alone and would not be expected to have any special interest in warning their neighbours that a predator is about. Since snakes are deaf, the rats cannot be whistling at them. Probably the benefit to the whistler is that alerting its neighbours deprives the predator of a meal that might persuade it to stay in the area.

Signalling to a predator that it has been spotted and that its chances of making a kill have vanished is taken one step further by bat-eared foxes and black-backed jackals, which mob larger predators with repeated, penetrating yaps that force the predator to move on by ruining its chances of catching anything at all in the area.

FIGHT

Even if they cannot bluff, outrun or outmanoeuvre pursuers, potential prey are still not lambs to the slaughter. A gemsbok is able to hold off spotted hyaenas and even lions as long as it can find a thorny bush to back into to protect its rear while it slashes with its horns at any attackers coming from the front. Giraffes and zebras use their hooves as defensive weapons; a kick can break a lion's jaw or fracture a hyaena's skull, and a giraffe's hooves have broken the back of more than one lion. Buffalo and blue wildebeest will go out of their way to harass carnivores, including those that are not hunting. Even when they are fleeing rather than attacking, a big herd

As well as fighting back when attacked, large herbivores will harass predators even when they are not hunting.

of large animals like buffalo is a serious threat to any predator that happens to get in its way. Animals the size of adult hippos, rhinos and elephants are simply too big and powerful for any carnivore except lions to tackle. In one case a bull hippo that was attacked by lions simply walked into the nearby water with them still clinging to his back! Young animals would be more vulnerable were they not defended by their formidable mothers.

A thick hide and formidable bulk are not the only ways of turning the tables on a predator. Thomson's gazelle ewes are successful eight times out of ten in driving off single spotted hyaenas that try to catch their lambs, admirable courage since a spotted hyaena is double a gazelle's weight. Suricates, banded mongooses and dwarf mongooses gang up against predators, making up what they lack in size by strength in numbers. A group of suricates or banded mongooses packed together and advancing in a bunch can put predators as big as jackals to flight by giving the impression of a single large and furious creature. Dwarf mongooses have a fixed battle formation for attacks on predators: the subadult males are in the vanguard with the rest of the subordinate adults in the rear. The alpha pair and the juveniles stay out of the action, though the alpha male may encourage his troops with a frenzy of scratching, scent marking and attacks on clumps of grass.

At 12–24 kg in weight a porcupine is big enough to provide an attractive meal even for a predator as big as a lion, and since its short, stumpy legs give it no chance of outrunning an attacker, natural selection has left it with a defensive armoury instead. From its shoulders to its tail the upper parts of a porcupine's body are covered by pliable spines up to 60 cm long and stiff, sharp quills up to 25 cm long, whose slightly flattened tips have cutting edges that improve penetration through skin. On the end of its tail is a rattle made of short, hollow, open-ended quills with flexible stalk-like bases.

While out foraging a porcupine makes little effort to conceal itself: sounds of its sniffing and gnawing, and its quills scraping against obstacles can be heard from a few metres away. If it hears or scents anything suspicious it freezes and listens intently, then carries on feeding or moves away quietly. If startled suddenly, or pursued and approached closely it erects its spines and quills and the crest of long, stiff hair on its head and shoulders, making itself look twice as big, and displaying its defensive weaponry. If the harassment persists the porcupine gives low-pitched roars, stamps its feet to rattle its quills together and rapidly vibrates its tail, which produces a hissing rattle from the bunch of short quills growing at its tip. It charges sideways or backwards at whatever is molesting it, trying to plant quills in its enemy. The speed of its rushes, with the momentum of the porcupine's body behind them, can drive the quills inches deep. The quills are only loosely rooted in the porcupine's skin, so they stay in the wound, and although they are not barbed they sometimes break off too near the skin for their victim to pull them out. If a quill stays in a wound it sets up an infection that can be fatal. Porcupines cannot shoot their quills, although the number of loose ones that lie around after an encounter may give the impression that they can.

A startled porcupine erects its spiny defensive rearguard. It carries quills only on its rear half, and because it cannot protect its head by rolling up like a hedgehog, it defends itself by erecting its quills, pointing them towards its attacker and manœuvring to keep its head away from danger.

Top: If attacked a hedgehog rolls into a spiny ball with its head tucked away from danger.

Above: A pangolin's scales are both armour and defensive weapons; their sharp edges can inflict serious wounds on an attacker.

On its first forays away from its den a baby porcupine is accompanied by an adult – its mother or father, or one of the other adults from its family group (*see* Chapter 5), who helps to protect it.

Although a porcupine's defence is spectacular, its effectiveness is limited if danger threatens from more than one direction. Only the back half of a porcupine is spiny, and if it is faced with more than one enemy its efforts to keep its quills aimed at one of them inevitably leave its head vulnerable to attack from the other. Unless the porcupine can find a burrow, or at least a hole big enough to get its head into, it is only a matter of time before one of its attackers lands a crippling blow to its head. In both the Kruger National Park and the former Kalahari Gemsbok National Park there are lions that specialise in hunting in pairs for porcupines.

A hedgehog's defence is less spectacular than a porcupine's, but probably more effective, not least because a hedgehog, weighing only about 300 g, provides a predator with hardly more than a mouthful, for which it is not worth getting pricked. When alarmed a hedgehog rolls itself up, and pulls the spiny skin of its back down over its head and legs, so that an attacker is faced with an immobile ball covered with interlocking prickles. The hedgehog may give a rhythmic puffing hiss, but otherwise it simply stays rolled up until the danger is past. The ball's immobility and lack of a head deprive many predators of the stimuli that trigger attack, and anything that does bite a hedgehog simply receives a mouthful of spines.

Pangolins are unique among mammals: their backs and flanks, the sides of their legs and the tops and sides of their tails are covered by thick, overlapping horny scales, so that they look more like reptiles than mammals. When threatened they roll up, with their heads and undersides out of reach, and if they are roughly handled they slide the tail sideways across the body, trying to catch their molester's snout or paws between the sharp edges of the scales, which can inflict deep cuts. A pangolin mother carries her single young on her back, and protects it by rolling up around it.

CHEMICAL WEAPONS

Mammals' defensive weaponry is not limited to teeth, horns, hooves, spines and armour: some of them have evolved chemical defences as well. The southern African mammal species that have the best-developed chemical weapons are the three mustelids (white-naped weasels, ratels and striped polecats) whose anal glands brew stinking concoctions of sulphur compounds and volatile acids that can choke and blind an attacker. To make them even more effective, a striped polecat can spray the contents of its anal glands. Members of all three species are strikingly patterned in black and white as a warning that they are not to be meddled with. A striped polecat will first flee from danger, then if it is hard pressed and cannot find refuge it fluffs up its long, silky fur to make itself look bigger, arches its tail over its back and turns away from its attacker while spitting and growling. As a last resort it squirts a stream of oily secretion from its anal glands. The secretion's odour is shockingly foul and it is a potent deterrent of mammalian predators, but unfortunately large numbers of striped polecats are killed on the roads when they stand their ground and spray at approaching vehicles. Ratels are rumoured to use their anal gland secretions to gas bees while they are raiding their hives.

That predators quite often have the tables turned on them by their intended prey – even when the prey appears to be completely outclassed in size, strength and weaponry – is due to the essential asymmetry of the predator-prey relationship: the prey is fighting for its life against a predator that is just working for its dinner. Being injured by a frantic opponent with nothing to lose is simply too high a price for a predator to pay for one meal, especially since an injury inflicted by one victim might rob the predator of the speed and agility that it needs to capture the next.

SELFISH HERDS

Significant protection from predators can result simply from being one of a crowd; and conversely, the threat of predation can be a driving force for the formation of groups. At the most basic level, group membership offers two benefits. If a predator catches prey one at a time, as most of them do, it will catch one from a group while the rest escape. The bigger the group, the smaller is any individual member's probability of being targeted for attack – what is known as the dilution effect. Animals will

The dilution effect – each individual springbok's chance of being the one that a predator picks gets smaller as the herd gets bigger.

also congregate simply because no potential prey wants to be the closest to a predator, and a group is a way of hiding behind other animals – the safest place is in the middle of a herd, which is where a blue wildebeest pursued by a predator flees to. Using other animals as cover produces what are known as selfish herds, and large herds of buffalo, blue and black wildebeest, and springbok fall into this category. Spinner dolphins scatter to feed at night and group during the day into what are probably also selfish herds, or in this case selfish schools.

The extra eyes, ears and noses in a group mean that there is a good chance that one animal will detect the approach of danger that the others are unaware of. Even if it does not give an alarm signal, its reaction to the threat will be enough to alert the rest of the group, which can then take evasive action. The individual who first spotted the danger also benefits; the sudden explosion of movement as its companions flee makes it less likely that the predator will make a kill. If a pursuing predator is going to take the first animal it catches up with, it pays not to be the last in the line. Inducing another animal to run behind may be the function of the white tail

The feeding impala would be perilously vulnerable to predators were it not for the watchful few that have lifted their heads and taken a break from grazing.

scuts and rump patches that are so widespread among antelope. A follower is no worse off than if it was fleeing on its own, and it benefits from a reduced risk of running into obstacles or ambushes.

The effect on predation risk of being in a selfish herd is demonstrated by the heavier predation suffered by territorial springbok rams in the former Kalahari Gemsbok National Park. The rams are often alone on their territories and their mortality rate from predation is double that of females in the breeding herds, who have the advantages of communal vigilance, the dilution effect, and the selfish herd.

Impalas and baboons are often together; for detecting danger the baboons' eyesight and tree-top sentinels complement the impalas' accute hearing and vision.

It is the threat of predation that drives vervet monkeys to live in groups that contain more than one male, while their forest-dwelling close relatives in Central and West Africa form groups with only one male. Because predators are more of a threat in the vervets' savanna woodland habitat, a dominant male is forced to tolerate the presence of potential competitors because they make a contribution to group vigilance. The alpha male still takes most of the burden on himself because he has a genetic stake in the survival of his young.

Selfish herds do not have to consist of animals of only one species. Baboons frequently associate with bushbuck and impala, while yellow mongooses, suricates and ground squirrels share burrows and respond to one another's alarm signals. There are also other benefits to these interspecific associations: bushbuck and impala feed on what their primate partners drop from the tree tops.

HEAT AND COLD

Extreme temperatures can be a major survival challenge, and the need to keep warm when it is cold, or cool when daytime shade temperatures reach 40 °C and the soil surface is at 75 °C has a major influence on behaviour. Cold, wet weather at the end of winter can kill kudu that are in poor condition after months of sparse feeding, and the distribution of nyala is limited by their sensitivity to cold. In the Willem Pretorius Nature Reserve buffalo changed their feeding behaviour so that they could shelter from the weather in riverine thickets, but because the sub-optimal habitat was also badly overgrazed 53 of them died from cold and hunger.

By far the best way of avoiding temperature extremes is to shelter in a burrow. In the Kalahari and Namib, air temperatures can reach 45 °C in summer, and fluctuate from -15 °C to 30 °C in winter. Soil surface temperatures are even more

extreme, reaching a maximum of 75 °C in summer. But 30 cm underground the annual range of soil temperatures is only about a third of the daily range in air temperature, providing a comfortable refuge from conditions on the surface. Humidity in a burrow is also higher than outside, helping to save loss of water through evaporation from the lungs.

Even a centimetre or two down, the temperature of the soil is significantly lower than on the surface, and mammals of many species from ground squirrels to spotted hyaenas make shallow scrapes and then lie in them with their bellies pressed to the cool subsurface soil. In addition, hyaenas urinate onto the patch that they lie on, and ground squirrels flick loose sand onto their backs while they are resting, and arch their tails over their backs as parasols while they forage.

The shade offered by vegetation or rocks can be an essential ally in the war against heat. Low-growing, dense bushes such as shepherd's trees or candle-pod acacias can hold pockets of shade that can be up to 20 °C cooler than the surrounding air.

The majority of animals avoid high daytime temperatures by becoming active around dusk and being back in shade or down a burrow before daybreak. Even those that are active during the day – such as elephants, giraffes, open plains grazers, baboons and suricates – take a siesta at midday.

Above: Ground squirrels can protect themselves from the sun's heat by using their tails as parasols.

Below: Warthogs, especially young ones, are perilously vulnerable to cold, and a burrow provides a warm night-time refuge.

Cold can be as serious a challenge as heat. A warthog's burrow is a refuge against night-time cold as much as against predators: if a warthog rests during the day it seeks the shade of a low bush. Neither warthog nor wild dog populations can survive unless they have burrows in which their young are protected from low temperatures. Wild dogs depend on burrows dug by members of other species, usually aardvarks. Warthogs can dig their own holes, and a warthog sow takes the additional precaution of providing a grass nest for insulation, and a recess in the back wall of the burrow in which the piglets will be safe and dry if water runs in and floods the floor. To escape the cold that settles into the dry river beds of the Kalahari on winter nights, springbok spend the nights in the dunes, even though the sandy surface there does not give them such good footing for flight from predators as the hard surface of the river beds.

Top: A cooling mud shower will help to remove parasites when the mud dries.

Above: Hot blood pumping through the veins at the back of an elephant's ear sheds excess body heat.

Bat-eared foxes forage at night in summer and during the day in winter, both in order to avoid low night-time temperatures and because their insect prey is also active during the day when it is cold. Similarly, aardwolves are nocturnal in summer, but winter cold keeps their harvester termite prey underground at night, obliging the aardwolves to become active during the day. Because food availability drops in winter aardwolves have to live off their reserves, and may lose a quarter of their body weight. To save energy they stay in their dens during cold spells and allow their body temperature to drop to 31 °C, relying on basking in the sun to warm up again.

All else being equal, large animals generate more metabolic heat than small ones, but in proportion to their bulk they have less surface area through which the heat can be lost. An elephant's huge body generates 4 kW of metabolic heat, and shedding it involves a number of special mechanisms. Its skin contains no sweat glands, but it is not waterproof, and so the elephant can lose heat by evaporating moisture through it. Wallowing in mud and water also helps to cool the body, as well as helping to remove external parasites such as ticks. Excess heat is also shed by rhythmically flapping the ears – there is a network of large blood vessels lying just underneath the thin skin on the back of each ear, and flapping them sets up air currents that cool the 12 litres per minute of blood that flows through these vessels by as much as 3 °C, and carry away 60 per cent of the elephant's heat load.

An elephant that has become overheated, perhaps because it has had to run during the heat of the day, or cannot find shade, may put its trunk into its mouth, suck up regurgitated water, and then spray it over its neck and behind its ears. If her calf gets too hot an elephant cow may spray it with water from her own stomach.

Hippos avoid high daytime temperatures by submerging in water. Spending the day in water also helps to avoid predators, and buoyancy saves a considerable amount of energy that would otherwise be needed to support their large bodies. As a result a hippo's food consumption is only 13 kg of grass a day, approximately a half of what would be expected for such a large animal.

Because the evaporation of water uses large amounts of heat, sweating is the most effective way for an animal to keep cool, but where water is in short supply, as it is at least seasonally over much of Africa, sweating may use too much of the precious liquid. The shortage of water grips most tightly in the desert and semi-desert of the Namib and Kalahari. Kalahari gemsbok spend 60 per cent of the daylight hours standing in shade, and graze at night when the moisture content of the forage is higher; even dead, dry grass can absorb 30 per cent of its weight in water when night-time temperatures fall. Springbok graze with their bright white heat-reflecting rumps towards the sun.

Above and below: Grazing at night and spending the day in water protects a hippo from the heat, and being buoyed up by the water cuts its energy requirements, and therefore its food intake, by about a half.

High daytime temperatures induce a state of stunned lethargy in the three big African carnivores – lions, leopards and spotted hyaenas. One of the reasons that cheetahs and wild dogs usually hunt during the day is that there is less chance then that their kills will be stolen. High temperatures may actually help wild dogs run down their prey: a fleeing antelope usually overheats faster than a pursuing wild dog, which can shed heat by panting.

Temperature regulation, feeding and the need to avoid predators such as caracals, black-backed jackals and black eagles all interact to determine the lifestyles of the rock dassies that live in colonies on outcrops of rock and boulders. The threat of predation keeps them close to the shelter of cracks and crevices between the rocks, and they rarely venture more than 15 metres from a bolthole. Suitable refuges are not very common, but those that do occur are usually big enough to accommodate more than one dassie, so they live in groups. Having several animals feeding on the plants close to shelter depletes the supply of palatable and nutritious vegetation, so most of a dassie's diet has to be poor-quality forage that is difficult to digest.

To add to its nutritional problems, feeding time has to be kept short to reduce the risk from predators. A dassie's digestive system is not particularly efficient at extracting the energy from food, and because it cannot improve its diet in either quality or quantity without exposing itself to predators, the dassie's only option is to reduce its energy expenditure. It saves energy by allowing its body temperature to fall, and by basking in the sun to warm itself up. On cold days dassies may not even emerge from their holes, but stay huddled together to share body heat.

Nests provide protection from both the elements and predators. The largest nests in relation to body size are the stick nests that Karoo bush rats build at the base of low-growing bushes. The nests can be as much as a metre high and contain more than 13 000 sticks, besides bits of bone, stones, mollusc shells and dry dung. Like a tree rat's nest, they are built from the accumulated leftovers of leafy twigs that the bush rats have bitten off and dragged back to the nest to be eaten.

PARASITES

Death from disease is insidious rather than spectacular, and animals' defences against it are nearly all physiological rather than behavioural. Lions sometimes eat green grass, probably to purge themselves of worms or hair balls. External parasites such as biting flies, ticks and fleas are a problem in themselves but are also the transmitters and vectors of a host of illnesses. Heavy infestations of ticks can suck so much blood that animals actually die from anaemia. In the sandy soil beneath dense

Occasionally, carnivores like this lioness swallow grass as a purgative.

shade trees, bloodsucking tampans can swarm in such numbers that animals prefer sparser shade where higher soil temperatures keep the tampans at bay.

Animals that use nests and dens are faced with a bloodsucking, disease-transmitting onslaught of fleas and lice that also use the den as a refuge and a breeding site. Movements between dens are frequently linked to declining local food supplies or attempts to evade predators, but in many cases the motivation for a move could instead, or also, be a desire to escape the attentions of parasitic vermin. A pangolin spends no longer than a week at one den before moving on to the next, and carnivore mothers often change dens every few days.

Elephants, rhinos, warthogs and buffalo bulls wallow luxuriantly in mud, which gives them a protective coating when it dries, and also makes it easier to rub ticks off. Rhinos especially have regular scratching spots on conveniently shaped rocks or tree stumps, and with regular use these rubbing posts develop a deep, rich polish. One of the very rare cases of tool use outside the primates is when an elephant picks up a stick with its trunk and uses it to scratch where its trunk alone cannot reach.

Top: An elephant shrew scratching unwelcome parasites from its fur.

Below: A plastering of dried mud helps white rhinos free themselves of bloodsucking ticks.

With the notable and somewhat puzzling exception of elephants, large herbivores have a symbiotic relationship with red-billed and yellow-billed oxpeckers, which feed on ticks that they remove from their mammal hosts. The oxpeckers also act as sentinels, and if they are disturbed they warn their host with a hiss.

Impala spend more time grooming than most antelope – they groom each other to remove ticks from areas that they cannot reach for themselves, and they are also the smallest antelope to host oxpeckers. An impala's lower incisor teeth are loose in their sockets, so that their tips spread apart when the animal grooms, and ticks in its coat are trapped between them and combed off. One of the penalties of old age is that as the teeth wear they become less effective, and old animals often carry huge tick burdens. This is a problem in the management of game ranches that do not have predators because the old animals shed massive numbers of ticks into the environment, which infest the rest of the population. If predators are present the old, decrepit antelope are hunted and eaten before their ticks can become a serious problem.

Above: A sprinkling of sand and a rub against a tree will dislodge external parasites from this elephant's hide.

Left: Adult male buffalo wallow in mud as protection against biting flies and to keep cool. Females and youngsters rarely wallow, suggesting that it also has social significance.

Opposite, top: With a hissing alarm call oxpeckers alert their hosts to lurking danger.

Opposite, bottom left: Impala groom each other in the spots that they cannot reach themselves.

Opposite, bottom right: Impala are the smallest antelope that regularly play host to oxpeckers.

Top, left and right: In almost every species of social mammal allogrooming has the dual functions of bonding groups together, and removing external parasites.

During the rutting season impala rams spend so much time fighting over territories, courting females and mating that they have no time left over for feeding, far less grooming. As a result they carry tick burdens that are six times as heavy as those on the ewes, and they play host to more oxpeckers than other impala.

The allogrooming that makes up such a large part of the interactions in almost all social groups not only cements the social structure, but is also a very effective means of removing external parasites, especially from the parts of the body that the animal cannot reach with its own teeth or hands.

FIRE

Each year in Africa south of the equator fires sweep across an average of 1,6 million square kilometres. Nearly all of southern Africa's vegetation is subject to periodic or sporadic fires, and considering the fury of a wind-driven fire in grassland, savanna woodland or fynbos, cases of mammals being killed by fires are remarkably rare. Small mammals simply disappear down their burrows – the heat from even the fiercest of blazes penetrates no more than a few centimetres into the soil. A radio-tracked genet survived a fire in its Cape fynbos habitat by sheltering between two boulders. The main danger comes not from the fire itself, but from its destruction of food supplies and cover that offers shelter from predators. The wind that drives a fire along also carries for kilometres the smell of smoke that warns of its approach, and larger animal species have the mobility to move out of a fire's path, onto rock outcrops or across rivers. Fires are likely to catch them only when fences block their escape routes.

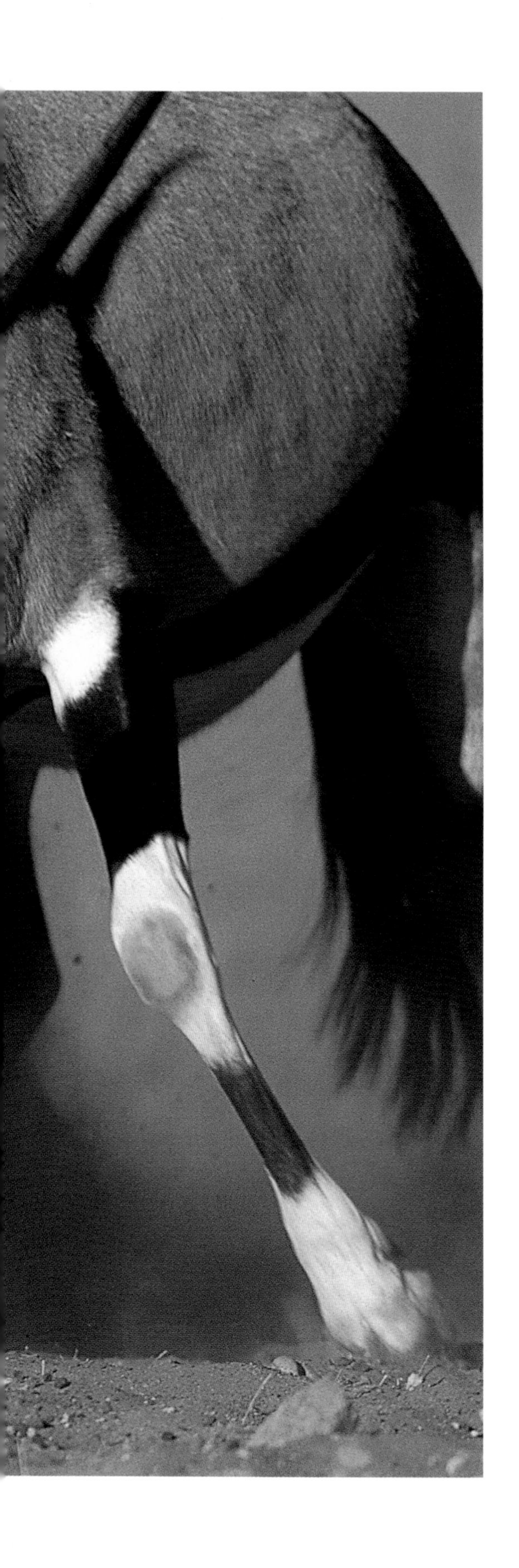

CHAPTER FIVE

SOCIAL LIFE

bonds and boundaries

TOGETHERNESS

SOCIALITY

ALL FOR ONE – GROUP DEFENCES

KIN SELECTION

CONFLICT

DOMINANCE

SPACE

GREGARIOUS LIFESTYLES ARE FOUND THROUGHOUT the animal kingdom. Flocks, shoals, swarms and colonies occur in thousands of different species, from protozoa upwards through invertebrates, fish, reptiles and birds. But it is among mammals that social organisation reaches its peak of complexity, flexibility and subtlety.

TOGETHERNESS

The conspicuous sociality of lions, elephants, monkeys and apes has been widely publicised, but there are other species, less obviously glamorous, whose members form groups just as intimately bonded and just as intriguing in their structures and dynamics. The complexity of social organisation among African mammals ranges from solitary wandering, through pairs and family groups and loose aggregations, to stable groups whose members collaborate to obtain food and avoid predators. It reaches a peak in the societies of dwarf mongooses, suricates, wild dogs and mole-rats, where individuals forfeit their own chances of reproduction to act as sentinels, guards, babysitters, nurses or food gatherers.

The smallest possible stable groups, mated pairs, are rather rare among mammals. Only about five per cent of mammal species have members that form pair bonds lasting longer than is required for a male to court a female and mate with her. Among the exceptions are Damara dik-dik, klipspringer, bat-eared fox, black-backed jackal and aardwolf, in all of which raising the young demands the efforts of both partners to bring them food or defend them against predators. Aardwolf parents work their guard duty in shifts, taking turns to forage or to stay at the burrow to defend their cubs against attacks from black-backed jackals.

The small antelope that live in pairs – klipspringer, Damara dik-dik, red duiker and blue duiker – have diets and habitats that make two the optimum group size to exploit and defend the resources in a territory. Since a highly desirable resource in any territory is an animal of the opposite sex, each territory is held by a mating pair, rather than by two animals of the same sex. In an exception that proves the rule (in the sense of testing it), all elephant shrews are territorial and monogamous, but this cannot be because males contribute to raising the young, since only four-toed elephant shrews and short-snouted elephant shrews live in pairs and in the other species males and females associate only for mating.

Previous pages: Gemsbok bulls battle for territory and access to breeding females.

Below: Klipspringers are quite common but their social structure – mated pairs – is rather rare among mammals.

SOCIALITY

An animal does not have to be social to live close to others, it simply has to be gregarious. True sociality is about interacting with others: even a solitary mammal has more social contact than a fish in a shoal or a bird in a migrating flock. A patch of fire-flushed grass, a tree full of ripe figs or a large carcass will attract animals from less-favoured areas, but each animal will mind its own business and what interactions there are will mostly be avoidance as personal spaces are infringed, or short squabbles over particularly succulent titbits. When the animals disperse they do so one by one; the group's size and membership are unstable and it lasts only as long as that particular clump of resources. These ephemeral, structureless, unstable groups are known as aggregations.

Not all social interactions are affiliative: territorial clashes that spread individuals apart and the xenophobic aggression that meets prospective immigrants to a group are just as much social interactions as the mutual grooming, sleeping together and babysitting that tie groups together. An established network of territorial neighbours can be thought of as a social group spread out in space.

The sociality of vervet monkeys is obvious, and it has been intensively researched. Sociality in other species may be more subtle, but it is no less intriguing.

Being a member of a group can improve an individual's food supply, reduce its chances of becoming a predator's next meal, help it to find a mate, and allow it to exploit localised resources such as a safe roosting site or a tree full of ripe fruit. But living in groups has costs as well as benefits – companions can be competitors as well as collaborators. A group may be more conspicuous than a single animal, attracting unwanted attention from predators, or confounding a hunt by alerting the prey. There may be competition for food: witness the scramble at a lion or spotted hyaena kill (but contrast the compulsive sharing of meat in a pack of wild dogs). Diseases transmit more readily between group members that share the continual intimate contact of allogrooming, mutual sniffing and licking, and sleeping together. The wild dog population of the Serengeti was wiped out completely by the wildfire spread of rabies and canine distemper.

In an apparent paradox, groups almost always have slightly more members than the optimum number that offers each member the widest margin of benefits over costs. The paradox is resolved by the huge benefits that a solitary individual gains by joining a group, compared to the small costs that having one extra member inflicts on the other group members – costs so small that it is not worth the trouble of chasing the newcomer away.

ALL FOR ONE – GROUP DEFENCES

If the extra vigilance provided by companions – whether organised as in a dwarf mongoose group, or incidental as in a herd of impala – fails to detect a predator before it launches its attack, being in a group can still yield the advantage of communal defence. In the face of a predator attack a family group of zebras bunches tightly and flees with its most vulnerable members – the youngest foals with their mothers – in front, and the stallion manoeuvring in the rearguard ready to use

Communal defence of vulnerable calves is one of the main reasons for elephants living in groups.

hooves and teeth on predators that venture too close. Buffalo herds will rally to the defence of their members, and can easily send lions scrambling up trees. A charge from a single adult elephant is more than enough to rout any predator, and elephants' close-knit sociality comes into play in a head-to-head contest of social systems when the two largest social super-predators (lions and spotted hyaenas) attempt to capture a baby elephant. The elephant-killing lions of Savuti choose their victims with care, targeting bulls whose companions can be relied on to desert them, rather than females who will rally invincibly to one another's defence.

If a pack of wild dogs on its kill is threatened by hyaenas some of the dogs will keep the scavengers at bay while the rest gobble down the meat. Once the carcass has been consumed the pack will abandon the scraps to the hyaenas. The dogs that fed will regurgitate meat for the guards.

If danger threatens a troop of baboons while they are near to a refuge they will scramble for cover, and devil take the hindmost, but if they are out in the open the adult males will act as a defensive rearguard. The formidable defensive abilities of adult male baboons are what keeps baboon meat off leopards' menus.

Perhaps the most direct and striking evidence for the importance of grouping to individual survival comes from cases of old, sick, crippled or disabled animals that are supported and cared for by their companions. Records of such cases range from campfire stories to well-documented studies by reputable field workers. Stories of blind buffalo managing to avoid predators by staying with their herds are quite common. There are well-documented cases of injured wild dogs, with broken or even amputated legs, being fed by their pack mates as long as they could keep up with the pack. Elephants, of course, are renowned for mutual aid, attacking predators and human hunters en masse, and using their tusks and trunks to lever fallen comrades back onto their feet. Dwarf mongooses feed and groom their sick companions, and even stay with them in their termite mound refuges until they recover or die, but what must be one of the most astonishing cases of care for an invalid is an observation from the Kalahari of one ratel feeding another whose back was broken.

KIN SELECTION

A golden thread running through the complex weave of interactions in mammal social groups is that youngsters are cared for by group members other than their parents. In black-backed jackals the pups of the previous year remain at the den and help to provision their baby siblings with food; elephant females act as nannies for one another's babies; brown hyaena cubs are provisioned with food by all the members of their clan, even though they were fathered by an itinerant stranger. A pride's lionesses suckle one another's cubs, and in a suricate, wild dog or dwarf mongoose group at least one adult stays on guard at the den as long as there are kittens or pups present. Because social groups are composed of close kin all these helpers are caring for their relatives, making as much of a contribution to their genetic heritage as they would have if they had bred themselves. As well as these enhancements of their genetic fitness, helpers may also benefit directly. A helper that stays in its parents' territory has a secure base from which to mount searches for

Young adult black-backed jackals remain with their families and help to raise their parents' next litter of pups.

a territory of its own; it picks up experience in the finer points of raising offspring, and, in jackals at least, it might occasionally take advantage of its position by hijacking a meal that was meant for the cubs.

Kin selection is one of the main reasons why sociality evolved, and even when social groups grow to include animals that are not close kin, ties of genetic relatedness can still be discerned in patterns of associations and interaction between individuals within the larger group. In vervet monkey troops, mothers and their offspring and grand-offspring, and even more distant relatives, support each other in disputes, and members of kinship groups interact more with each other than with the rest of the troop. Threats and attacks are influenced by kinship, and even by the interactions of one monkey with another's relatives. If a female vervet is displaced from food she may vent her feelings by attacking her displacer's relatives. Even within the anonymity of a buffalo herd, social subunits of 15–20 animals travel and graze together, and within these subunits family groups of female relatives tend to bunch together. Elephant herds are associations of family units of a cow with her calves.

Although groups are built on an evolutionary foundation of kin selection and reciprocal altruism (*see* Chapter 1) it is allogrooming that provides both glue for the social structure and grease for the social mechanism. Social rodents such as Angoni vlei rats, tree squirrels and ground squirrels, carnivores from suricates to lions, and herbivores such as impalas and zebras allogroom by nibbling and licking. Primates use their hands, working systematically through the partner's fur and removing flakes of dead skin, parasites and salt crystals.

Social mongooses, rock dassies and tree squirrels develop a unique group odour because they sleep together, frequently groom one another, and even scent-mark each other. Absences as short as a few hours can affect either the odour itself or the group's recognition of it, and after such temporary absences group members must be cautious as they approach the group if they are not to be mistaken for a stranger.

Below, left: Allogrooming between vervet monkeys is a family affair.

Below, right: For dominant male baboons one of the privileges of rank is to be the focus of allogrooming.

In their social interactions, choices of coalition partners, dispersal and anti-predator strategies, animals do not calculate their degree of relatedness to their companions in order to work out how they should behave towards them, any more than they study nutrition when deciding what to eat. They merely behave as if they are doing the calculations. The rule of thumb appears to be that animals treat those they grow up with as if they are closely related – which in nearly all cases, of course, they would be.

Ground squirrels nibble and lick the fur of their colony mates. This allogrooming acts as a social glue that reinforces bonds between group members.

CONFLICT

Direct conflict between animals of the same species is just one aspect of competition for scarce resources, be they food, safe refuges or females on heat.

Despite the numerous occasions on which animals have both motive and opportunity for it, fighting is rather rare, full-scale battles rarer still, and fights to the death so rare that they call for special notice. Ritualisation, parading and posturing are very common as a preliminary to combat, or apparently as a substitute for it, and most conflicts are settled by threats or cursory tests of strength in which there is not much chance that either opponent will be injured.

Even the battles themselves may appear to be ritualised, with blows being aimed where they will do least damage – nonetheless, the importance of ritualised 'fighting by the rules' in combat between members of the same species has probably been exaggerated. That neither combatant goes for a reckless all-out attack is more likely to be because both are prudent enough to want to avoid being injured, while still trying their best to inflict significant damage on the opponent. A kudu bull who carefully locks his horns against a rival's is not sportingly ensuring that he cannot injure his challenger, but making as certain as he can that the challenger cannot injure him. When two seals fight, most of the bites land on their protected necks, because that is the only part that can be attacked without leaving an opening for dangerous retaliatory bites to the rump and flippers.

The conflicts in which apparent ritualisation is most striking are fights between giraffe bulls. The outcomes of these battles determine rank in the local dominance hierarchy, and with it access to females that are ripe for mating and impregnation. A giraffe's weapon against other giraffes is its head, which it swings like a mediaeval knight's mace, landing thudding blows on its opponent's body and legs. Like a mace, a giraffe bull's head is studded with an assortment of knobs, and a pair of short, blunt horns grow from the roof of its skull, which at its full development is solid bone 10 cm thick. The length of a giraffe's neck gives its movements a deceptively deliberate quality, but when a blow lands it can break the opponent's ribs, neck or leg, or cause internal haemorrhages that are ultimately fatal. The sound of the impacts can be heard hundreds of metres away. A captive giraffe bull in Frankfurt zoo struck the eland bull that shared its enclosure a single blow with its head. The eland was knocked off its feet and killed.

Each combatant rides his opponent's blows by jumping slightly at the moment of impact. It is because they cannot jump and swing at the same time that they give the impression of taking turns to give and receive blows. During the pauses in their mutual battering the two manoeuvre for position, shoving against each other, trying to trip one another, or at least deprive the opponent of the firm footing that he needs to drive home a telling blow or to ride one out. What appears to be ritualisation, a gentlemanly fight by some unwritten version of boxing's Queensberry rules, is in reality a complex pattern of attack and defence, feint and parry, probing for the opportunity to land a telling blow without leaving an opening for retaliation.

The purpose of fighting is to win, and if the opponent persists until he is injured, crippled or killed instead of fleeing when the battle began to go badly, then so much the worse for him. Animal conflicts rarely lead to fatalities – losers break and run as soon as they begin to get the worse of an exchange.

Postures that display weapons such as horns and teeth, or emphasise the size of the body – standing broadside on, fluffing up the hair, hunching the shoulders and standing tall – are usually threats. Crouching, sleeking the hair, and de-emphasising weapons are submissive signals. Threat displays emphasise and even exaggerate those features that contribute to fighting ability: for a carnivore to bare its teeth is a universal threat signal that is made offensive or defensive by the exact positioning of lips, ears and eyes. A mature kudu bull's mane and an eland bull's dewlap make their necks look even thicker than they are.

Threats can be offensive or defensive. Offensive threat means something like 'if you do not back down I am going to attack you', while defensive threat is 'if you attack me I will defend myself with violence'. Defensive threat is not the same as submission, which sends the message 'I dare not retaliate'.

For one combatant to withdraw after the displays of size, strength and weaponry in the build-up to a fight, or the ritualised preliminaries to all-out conflict, yields benefits to both parties; the weaker of the two avoids a thrashing, and the stronger gains by winning without having had to fight. Both benefit by avoiding the time and energy cost and risks of injury of battling it out.

Bottom: The atmosphere of stately ritual in fights between giraffe bulls is an illusion conjured by their complex fighting style in which jockeying for position is as important as striking a blow.

Opposite: An elephant's threat pose emphasises size and weaponry.

A warning snarl from a lioness shows off her fearsome canine teeth.

If faced by an opponent whose displays show that he would be hard to beat, a challenger would do well to find himself a weaker opponent. It could be that displays by resource holders serve to convince challengers not so much that the holder cannot be beaten, but that an easier conquest can be had elsewhere.

DOMINANCE

In all mammal social groups, and often among solitary animals who meet each other regularly, disputes over resources such as food and comfortable resting sites are far more commonly settled by one animal deferring to the other than by outright conflict. Such an arrangement is almost essential to group cohesion – no society could survive if every dispute was settled by a duel. The individual who is deferred to, and who therefore has privileged access to valuable resources, is said to dominate the other, and the network of similar interactions that weaves through a group is known as a dominance hierarchy.

Once a dominance relationship has been established, the signals of rank that maintain it can be very subtle indeed. In a troop of baboons or vervet monkeys, a direct approach by a high-ranking male is all that it takes to send subordinates slinking out of his path. A dominant buffalo bull can displace another from a shady spot or a mud wallow by a shake of his head. Frequent squabbles and fights are a sign that a stable dominance-subordination relationship has not been established – if it had, the subordinate would simply defer to the dominant, relinquishing whatever resources were in dispute.

Like territoriality, dominance in groups may concern some resources and not others. Among yellow mongooses, for example, high rank confers preferential access to feeding sites, but a subordinate will not relinquish food that it has already found.

Rank may depend on sex or age, independent of individual identity. In a herd of buffalo, all the bulls are dominant to all the cows, regardless of individual identity, and within each sex there are separate hierarchies in which rank is determined by fighting ability, which in turn depends on size and strength. Any buffalo can decide whether to defer to another simply by seeing which sex it belongs to and how big it is.

In a clan of spotted hyaenas the males are subordinate to the larger, heavier, more aggressive females and to cubs; among the females there is a dominance hierarchy based on size and savagery, and among the males immigrants outrank residents. Privileged access to food and the best dens enables a high-ranking female to double her production of cubs, and it is this potent selective advantage that has driven the evolution of extreme aggressiveness in female spotted hyaenas. This aggressiveness is a consequence of masculinization through exposure of female hyaenas to male hormones while they are still in their mothers' womb. Masculinization of behaviour brings with it a suite of other male characters, such as a fat-filled pseudoscrotum and an erectile clitoris that is as large as a male's penis. These bizarre genitals play a leading role in the hyaena greeting ceremony (Chapter 6), even though they arose as an evolutionary by-product. The cost of being masculinized include a high rate of still-births, and fatal battles between new-born twin sisters (Chapter 2), but these are more than counterbalanced by the benefits of the high rank that aggressiveness confers.

In suricates, wild dogs and dwarf mongooses, adults defer to juveniles in competition over food, and the juveniles enjoy a high rank that they would be quite unable to enforce if challenged. Wild dog juveniles do not hesitate to snap at adults that try to feed before the juveniles have finished, but the enforcers of the social code are the dominant breeding pair.

In elephant family groups and the more richly structured carnivore and primate societies, dominance hierarchies – and indeed the whole social structure – are based on individual recognition. Each group member learns its place by remembering past encounters. Because the subordinates are conditioned to defer to a particular individual, it can retain its dominant position even when it is past the peak of physical condition that originally allowed it to climb in rank.

By far the most common situation is for an animal's rank to depend on its fighting ability and for it to have to enforce its status every so often by giving a painful lesson to a challenger from the lower ranks. When the dominant's fighting abilities wane it begins to fail these tests and so drops down the hierarchy. A dominance hierarchy tends to be self-reinforcing. A high-ranking animal eats better, sleeps more comfortably and suffers less stress than those below it, and so stays in better physical condition than they do. An offspring of a dominant female has a head start in life because its mother is better able to feed it, and in many social systems high rank is bequeathed by mothers to their daughters.

It is not only dominance-subordination relationships that are ordered hierarchically. Equivalent asymmetries may appear in other interactions such as allogrooming, or who greets whom, or which individuals or age-classes are most diligent in disputes with territorial neighbours. Very often these other relationships show some kind of link to dominance – for example, high-ranking male baboons are a focus for allogrooming by others, but very rarely initiate any allogrooming themselves.

A dominant male baboon intimidates rivals with a threat yawn.

Females with young babies tend to stay close to high-ranking males.

Within a baboon troop the adult males form a dominance hierarchy that is established by fighting and maintained by staring, displays of dental weaponry, and chasing. A submissive animal opens its mouth only slightly, and pulls its lips back in a grin, while a dominant one gives display yawns with its lips lifted to expose its canine teeth. A dispute between two males often sets off a chain of squabbles, chases and fights in the rest of the troop. The highest ranking male is the most aggressive member of the troop, and also its most assiduous defender; other troop members present their rumps more often to him than to each other, and females whose babies are still in their black natal coats tend to stay near him. Low-ranking males occasionally form coalitions with one or two partners and gang up against dominant males. Males play with and groom infants, sometimes as a way of defusing aggression from dominant animals. A high-ranking male gets first choice of feeding site, and other adults keep clear of him, but he will tolerate approaches by the babies of his female social partners.

There is also a hierarchy among a troop's females, but it is fluid because each female's status depends on her reproductive condition, which male she is with, and whether she has a baby. A female baboon inherits her mother's status and so each troop contains a number of matrilines that are hierarchically arranged relative to one another. In social conflicts females support their offspring and grand-offspring, and males support their offspring.

In other species, rank in a dominance hierarchy is likewise often linked to reproductive condition, and can change quite sharply with reproductive cycles. Buffalo cows rise in status when they have a calf at foot, and are then more likely to be in the centre of the herd, towards the front, the position that offers the best compromise between food availability and safety from predators. Female vervet monkeys go up in social rank when they have a baby. Full-grown African elephant bulls go through periodic shifts in reproductive activity and dominance status in a phenomenon known as musth.

When a bull is in musth he holds his head high and walks with a distinctive rhythmic swagger. He becomes extremely aggressive towards other elephant bulls, and is assiduously avoided by males who are not in musth. The temporal glands on each side of his face, about halfway between the eye and the ear, swell and discharge a sticky secretion that stains his cheeks with a broad, dark streak. He rubs the glands against trees, and sometimes fragments of bark or broken twigs become stuck in the glands' openings. Instead of urinating with his penis extended he does so in a continual dribble with his penis retracted into its sheath. The sheath and his hind legs become stained with urine, which develops a strong odour that is unpleasant to humans but apparently attractive to elephant cows, since they will follow the trail of dribbled urine that a musth bull leaves behind him. Bulls in musth are very attractive to females, who prefer them as mates (*see* Chapter 7).

Bulls that are not in musth always back down when challenged by a musth bull, but on the rare occasions when two musth bulls meet, the confrontation is a spectacle of uninhibited savagery. The combatants charge head-on, slam into one another and wrestle with their trunks and tusks. If he can, the bull that is getting the worse of the exchange will break and run, but he risks being tusked in the rump or flank as he turns to flee, and the winner will chase him for as much as three kilometres, even in the midday heat, and will tusk him if he catches him. Losers can be killed outright in such battles, and at least one case is known of a victor dying later of his wounds. In small protected areas such as the Addo National Park and the Tembe Elephant Reserve in South Africa attacks by musth bulls on other elephants who cannot get away because of fences are a significant management problem.

Fights between musth bulls are real tests of strength and motivation, and are probably why musth is such a reliable indicator of an elephant bull's calibre. Any bull that 'cheated' by going into musth when he was not at the peak of physical condition would run a serious risk of being beaten and tusked to death by a genuine musth bull who really was in his prime.

In hierarchical systems, the low-ranking animals have to make do with poorer habitats, fewer resources, less secure resting sites and no access to females. But they are not playing an understudy role in an organised scheme of things that somehow benefits the species by reserving the cream of the resources for the best specimens, and ensuring that only the healthiest and strongest individuals breed. What they are really doing is biding their time. When they have grown in strength and experience they will take advantage of any weakness in the higher ranks; perhaps a territory holder grown old and feeble, injured in a fight with another challenger, or killed by a predator.

Both parties in a dominance-subordination relationship come away from each interaction better off than they would have if the relationship had not existed. For a feeble, subordinate animal, deferring to a dominant one is better than being attacked and losing anyway. For a vigorous, high-ranking animal pulling rank saves energy, and is less risky than having to fight at every encounter. There are certainly benefits to the group: less fighting leaves more time for social bonding and vigilance against predators; but these are fringe benefits, not the selective driving force behind the evolution of dominance.

An elephant bull in musth signals his condition by his posture, the secretions of his temporal glands that make dark smears on his cheeks, and the dribbled urine that stains his hind legs.

SPACE

A home range is simply the area in which an animal goes about its normal business. It may be a definite area in which the animal spends all its time, with regular movements from one place to another along established pathways, or a diffuse region through which the animal wanders with no set pattern. Of course, space itself is not particularly useful; the real resources are the food, resting sites, females or whatever that are scattered through it. In fact, too much space is a liability – the best home ranges are those where the most resources are crammed into the smallest area.

As resources become scarcer and more scattered, home ranges have to become larger. In the semi-arid Kalahari, spotted hyaenas scratch for a living over areas of 1 000 km^2; in the far more productive Kruger National Park they can live comfortably on 150 km^2. Klipspringer pairs hold territories of 15 ha in the Karoo and 49 ha in the more arid and less productive Northern Cape.

A territory is an area in which residents have privileged access to the resources. These are nearly always food or females, or the food that attracts females. An animal, or a group, will hold a territory only when it can be defended economically – in other words, when the benefits outweigh the costs. If resources are scarce, widely scattered or unreliable, a territory that is big enough to contain sufficient of them will be too big to be defendable. At the other end of the scale are resources that are so abundant that holding a territory does not increase their availability, and in these cases too we would not expect to see territorial behaviour. Reedbuck live in pairs or family groups. In the Kruger National Park, where the food supply is reasonably rich and not markedly seasonal, the rams hold territories throughout the year. In the Underberg and Highlands regions of KwaZulu-Natal resources are scarce in the winter, and males hold territories only during the April to August mating season. Around St Lucia the feeding is so rich that territoriality has no pay-offs, and the rams form dominance hierarchies instead.

Below and opposite: The stakes are high for female leopards fighting savagely over territory; they need priority of access to an area's resources in order to raise their cubs.

The resources that a territory contains increase in proportion to its area, but the length of the perimeter that has to be demarcated and defended increases only as fast as the square root of the area. A territory with double the area and double the resources has only 1,4 times as long a boundary, and a territory with four times the resources has only double the perimeter. If four solitary territory holders were to amalgamate their holdings and defend them as a group, each animal's defensive duties would be cut by half, leaving more time for feeding, courtship, avoiding predators, and other behaviour.

Territory size and group size need not necessarily always be related. In some animal populations the sizes of groups are set by the richness of the resource patches – the more food each patch provides, the larger the groups can be – and the sizes of territories depend on how many patches are needed to provide a reliable resource base, and how widely they are scattered.

A typical cat's social life and spatial organisation would show solitary females who hold territories in order to defend access to food resources, and solitary males whose territories overlap those of several females in order to keep other males away from them. In build and hunting behaviour cheetahs are far from being typical cats, and their spatial organisation also has some unusual features. The females are typically feline in that they are usually solitary; their home ranges of a few hundred square kilometres overlap with those of other females but they avoid meeting by using their shared areas at different times. Males hold territories in areas that are popular with females, but unlike most cats they form coalitions, usually with two or three members, to share the duties of territorial defence. Coalitions are usually groups of brothers: within the coalition they share matings, but their territory and the females it contains are fiercely defended against intruding males in battles that are quite commonly fatal. Defeated opponents may even be eaten.

Territories and home ranges can be unexpectedly stable. One klipspringer ram held an area for eight years; a nursery herd of roan antelope occupied a home range for 30 years, and three lion prides in the Serengeti have lived in the same home ranges for more than 20 years.

Cape fur seal bulls gather harems of females that they defend from other males. Once a male has established a territory he guards it continuously for about six weeks, living off his stored blubber.

Long tenure enhances one of the major advantages that an animal gains from having a territory or a regularly used home range – namely an intimate familiarity with the area it lives in; knowing the twists and turns of its pathways, the location of dens and bolt-holes, the shortest routes between food and water, and where the sun shines warmly on a cold winter's morning. Wherever it happens to be in its home range, an aardvark must know where the nearest of its many burrows is dug, because it makes straight for it when danger threatens. The benefits in terms of efficient use of resources and safety from rivals and predators, and the inevitable investment in time without which the features of an area cannot be learned, generate an inescapable asymmetry whenever the holder of territory is challenged by an interloper. Because the resident already knows its area, and the interloper does not, the resident has more to lose by being driven out of the territory than the interloper has to gain by taking it over. As a result, the resident will be prepared to fight harder, and to escalate the conflict to more serious levels than the interloper would be. Only if the resident is markedly weaker than the intruder is a takeover likely to get beyond the testing stage. The exceptions to this rule, that prove it in both senses of the word, are territories like those of impala rams or fur seals in which the only resources are females to be mated with, where intimate familiarity with the locality is unnecessary, and where the temporary nature of the territory would make it impossible. On these territories, which are hardly more than mating grounds, pitched battles and takeovers are common occurrences, and success at holding a territory or in displacing a resident depends on strength and skill in battle, not on squatter's rights.

Territory is social dominance with a spatial component; a territory is an area where an animal dominates others that could dominate it elsewhere. White rhino bulls provide the classic example. After they mature at about 12 years old they establish territories which they demarcate with dung middens (*see* Chapter 6) and by spraying bushes with urine. A male intruder or a subordinate male group member will be tolerated as long as he behaves submissively by standing with ears back and tail up while roaring and squealing. The territory holder may approach a subordinate and briefly push horn to horn, or may simply ignore him. In turn, a bull that leaves his territory will be allowed to pass unmolested as long as he behaves submissively and shows no interest in females. While off his territory a bull will not spray urine or kick dung, but as soon as he is back on his home ground these behaviours reappear. Cows move freely in and out of male territories, but when a cow is on heat a territorial bull will try to keep her inside his area. Because only territory holders get a chance to mate, rights over territory – and the access to breeding females that it confers – are decided by outright combat. The preliminaries to fighting are deliberate approaches and charges. Battles escalate from horn fencing, through jabbing with the horn, to heavy shoulder ramming and hooking at the opponent's body. Serious injuries and deaths most often occur when there is an oestrous female as the immediate prize.

That they are territorial is one of the many puzzling things about elephant shrews. Presumably the resource they are defending is food, because most species are solitary as well as territorial, but the territories are huge for such small animals: a 40-gram round-eared elephant shrew may have a territory as big as the area held by a band of ten dwarf mongooses that weigh 250 grams each.

The more valuable the resource that the territory holder enjoys privileged access to, the greater the investment in territory establishment, maintenance and defence that will be worth while. Consequently the most conspicuously costly territoriality is found among rutting impala rams and Cape fur seal bulls in breeding colonies, because the resource that their territories contain is reproductive females.

Interactions between territorial neighbours on their mutual boundaries are very often highly ritualised. In waterbuck, for example, when territory neighbours who know each other meet, one stands broadside to the other with his chin tucked in, his head high and his horns tilted towards his neighbour, then lifts his head to the horizontal. The neighbour makes the same movements, and the two repeat the performance as they move closer together and come to stand head to tail. They then swing round until they are facing each other, swish their tails and make attacking feints and blocks. The end of the encounter usually comes when the bulls take turns to advance a short distance while the other retreats, shaking his head from side to side. They then move apart to graze without any physical contact having occurred.

White rhino bulls stand horn to horn, back away, wipe their horns on the ground, and advance and touch horns again in a repetitive ritual that can go on for over an hour. Finally they back away, and turn and walk off, spray-urinating and scraping the ground with their hind feet as they go.

These ritualised encounters between neighbours are exactly what is expected. Since they both hold territories already they must both be battle-tested veterans with similar fighting abilities, who would inflict, and suffer, serious damage in a fight. For an animal that already holds a territory, taking over another one yields no

Rhinos scent-mark with urine (top) and by defecating in middens (above). Males spray backwards between their legs into bushes while standing still.

advantage because it simply expands his area beyond what it is possible for him to defend economically. Territorial battles are usually not between established residents but between a territory holder who has a lot to lose, and a landless intruder who has almost as much to gain.

In a long list of species that includes white rhinos, waterbuck, blue and black wildebeest, reedbuck, red lechwe and springbok, males are territorial while females live singly or in herds with home ranges that encompass the territories of several males. When the females move through a male's territory he checks their reproductive condition by sniffing their urine and genitals (*see* Chapter 7) and courts and mates with any that are receptive. He tries to keep females within his territory as long as possible, by herding them away from the border when they try to leave. In all these species the males that are at the peak of condition control the richest territories in the best habitat, containing far richer food resources than the male requires for his own needs. But the food supply is not the primary reason why the male holds the territory; the resource that he is really concerned with is the females that also use the area, and territories with richer food supplies attract more females and hold them for longer.

Above and below: Adult springbok rams expel other males from their breeding territory, and herd females into them.

Springbok rams stay on their territories even when conditions have deteriorated so much that all the other springbok have deserted the area. Their payoff is that they are established and ready to begin courting and mating when improved conditions attract the breeding herds back again. Rams with less persistence have to use up valuable time in fighting for territories before they can get down to the business of attracting females.

Once females have formed groups, for whatever reason, they become a resource that is valuable to males in its own right, and social systems can develop in which males monopolise access to female groups. These social systems are found in mammals as diverse as hippos, lions, rock dassies and roan antelope.

To escape the daytime heat, hippo cows crowd together in rivers and along the shoreline of lakes, ideally where gently sloping banks enable them to climb out on their nightly grazing forays and the water is deep enough for them to submerge, but shallow enough for them to stand on the bottom with their eyes and nostrils exposed. Once they are on land they scatter over a wide area, travelling several kilometres, if they have to, to reach grazing. It is the daytime concentrations that fall under the control of a territorial bull, who monopolises access to his harem of females by evicting all other adult males from the stretch of water that his females use.

A territorial bull displays at intruders by charging, head shaking, roaring and grunting, threat yawns which display his dental weaponry, scooping water with his lower jaw, and showering dung and urine over them. He will attack any intruder that does not withdraw, and battles over territory are uncompromisingly savage. A hippo's front teeth, the incisors and canines, are not used at all for feeding; instead, they are specialised as combat weapons. A mature bull's lower canines stand 30–50 cm above the gums, and the cutting edges of their triangular tips are kept sharp by abrasion against their opposite number in the upper jaw. They penetrate even the 6 cm thick hide on a hippo bull's neck, and inflict dreadful stab and slash wounds, which can be fatal.

A hippo bull's territory is one of the few among mammals from which all competitors are completely excluded. There is no equivalent among hippos of white rhino bulls' tolerance of trespassers who behave submissively. Beaten opponents are driven from the water by repeated savage attacks, and if there is no other water body within walking distance expulsion is effectively a death sentence.

On a less extravagant and spectacular scale, rock dassies have a social system remarkably similar to that among hippos. Groups of up to 17 females congregate on rock outcrops where crevices offer shelter from night-time cold and refuge from predators such as black eagles, leopards and caracals. A single mature male holds territorial sway over the females' home range, perhaps with another male hanging around on the periphery. He keeps other adult males out, and drives off the group's male offspring when they are about 15 months old by ferocious attacks with his long, sharp incisor teeth. Just as in hippos, a dassie's front teeth are specialised as weapons, with triangular tips honed along their edges by wearing against the opposite pair.

A group of females does not have to stay in one place for a male to establish exclusive rights over it. Roan antelope have undefended home ranges, but a bull chases all other males out of a 300–500 metre wide exclusion zone around the breeding herd that is in his home range. The exclusion zone is not a territory, because it moves around with the breeding herd.

Hippo bulls fight savagely over territories and the females that live there.

Above: A young male lion killed during a pride take-over.

Opposite, top: This is just one of the one to two hundred copulations that this lion pair will perform during the female's receptive period.

Opposite, bottom: While a lioness is on heat her pride males usually stay in close attendance.

LIONS

Lions have among the most complex and intriguing social and mating systems, and the most intensively studied, of all the African predators. In bare outline, the females live in stable social groups of up to 12 members, with hardly any exchange of members between groups; they grow up, live and die in the group that they were born in. Typically a female group is attended by a group of males, who remain with them for a period of several months or a few years until they are replaced by another male group. Together the males and females make up a pride.

The females within a pride are closely related, on average slightly more closely than second cousins. This gives them a genetic stake in each other's reproduction, which surfaces most obviously in their suckling of one another's cubs. Females show only a weak favouritism for their own cubs, mainly by terminating suckling bouts soon after their own cubs have filled themselves, even if other cubs want to carry on suckling.

Single males manage to hold onto their prides for an average of about a year, coalitions of two for a year and a half, coalitions of three for three and a half years, and of four to six for four to eight years. The longest known tenures are about ten years. Replacement of one male group by another is invariably a violent upheaval. On average the new males can look forward to a tenure of about three and a half years, just long enough for the lionesses to raise two litters of cubs, provided that they start breeding straight away. So that the lionesses do not waste any of the new males' breeding cycle by completing the raising of the ousted males' cubs, the new males kill all the cubs under nine months old and drive out any older cubs and subadults, including females that are too young to breed. What remains under the males' control is a group of sexually mature lionesses who have just lost their cubs, or are pregnant. Loss of her cubs can bring a lioness back into breeding condition within a few days, but mating may begin almost immediately as the females try to use sex to distract the males from infanticide. Because the females prefer to be under a strong coalition whose long tenure will delay the next infanticidal take-over, mating activity can go on for three months without the females conceiving, during which time the new males have to prove that they can defend their position against challenges from other males.

Once they become fertile again the lionesses tend to breed in synchrony. Having all the pride's cubs about the same age makes their shared upbringing easier, and it ensures that when the male cubs disperse they can do so as a coalition of litter siblings or half-siblings that will have a better chance than a single male of taking over and holding on to a pride. Males that have to disperse alone generally make the best of a bad job by joining up with another loner to form a coalition pair.

In complete contrast to the savagery and bloodshed of pride take-overs, there is strikingly little competition between coalition partners for opportunities to mate. There are a number of interconnected reasons for this sexual egalitarianism. Lions mate two to four times an hour for up to three days. The male's libido wanes as the number of matings steadily climbs, and when her first partner can no longer be aroused, a lioness on heat will move on to the next. The function of her insatiable sexual appetite may be to reduce competition between her pride males: with so many mating opportunities available, they are simply not worth fighting over. And since each coupling has only a very small chance of leading to conception there is no selective pressure on males to try to monopolise matings at the expense of their coalition partners – sooner or later they will get a chance to mate when the lioness has worn out her current partner, and any one or more of the matings could lead to conception. In addition, coalition partners are often closely related – full brothers, half-brothers or first cousins – and having genes in common further reduces the evolutionary pressure towards competing for matings. Co-operation, even if it is motivated by genetic self-interest, is the evolutionary glue that holds groups together.

MONGOOSES

Although they have received much attention from researchers, the social systems of lions and elephants are not, in fact, exceptionally complex by mammalian standards. Many of their subtleties have been brought to light only as a result of the long periods of intense research effort that have been focused on them. If the same resources were poured into field work on other social species, it is more than likely that similar intriguing features would emerge.

Yellow mongooses are on one of the middle steps on the ladder of social organisation. They live in colonies with up to 13 members – usually 2–3 adults and their young – who share a warren but forage alone during the day. A colony's females come into heat at different times and so a single dominant male can monopolise matings by servicing them one at a time. The subordinate males visit the females of neighbouring groups. Females give birth within a few days of each other, and they suckle each other's babies (though each female prefers to suckle her own offspring). Other colony members babysit when the mothers are out foraging, and they also bring food to the den for the youngsters, the only social mongooses that do so.

If degree of sociality is measured only by group size, the carnivore species that ranks the highest is the banded mongoose. Their usual pack strength is about a dozen adults, more or less the same as a large lion pride, and they commonly form groups of about 30, which match spotted hyaena clans for numbers, but the largest pack ever recorded had 75 members, which is more than any other social carnivore. Several females come on heat at the same time, which spreads mating opportunities among the males, because no single male can attend to all the females at once. Females litter down together and suckle one another's offspring indiscriminately, a social step up from yellow mongoose mothers, who favour their own offspring. They forage in groups, which keep in contact with continual high-pitched twittering calls. The whole group responds to its members' alarm calls, either freezing, standing up to locate the danger and then slipping quietly away, or diving for cover. Youngsters are protected by the adults, and there is one case on record, from East Africa, in which the pack's alpha male climbed a tree and rescued a pack member that had been captured by a martial eagle! Nevertheless,

Dwarf mongoose groups are extended families. Their intense sociality has evolved by kin selection.

banded mongooses do not have the sophisticated sentinel systems that are found in dwarf mongooses and suricates, which occupy the twin peaks of African carnivore social organisation.

A suricate sentinel keeps watch for predators that threaten its group.

Dwarf mongooses are the smallest of African carnivores – with a body weight of 210-340 g, they are about the size of a one-month-old domestic kitten. They fall within the prey size range of small carnivores, monitor lizards, large snakes and all predatory birds from the size of pale chanting goshawks upwards. They forage only in daylight, and often over open ground, which makes them extremely vulnerable to predators, and utterly dependent for survival on the system of sentinels which is the keystone of their social organisation. From the moment that a mongoose group emerges from the warm refuge of the termite mound where it spent the night, it has at least one of its members watching for danger. The first mongoose to emerge as the rising sun touches the mound stands on its hind legs, with its tail as the third leg of a tripod, and checks all around for predators. Its companions emerge only when it gives a short 'peep' as an all-clear signal. Even during the intense session of socialisation and grooming that gets the group's day started there is a sentinel on duty, and when they move out to forage they are preceded by the next sentinel, who climbs up onto a vantage point along the group's line of travel, and takes over responsibility from the one near the mound. And so it goes throughout the day, with sentinels taking turns at duty spells of 15–20 minutes, and only coming off duty when the next one is up and calling the all-clear. Most predatory attacks come from behind the group, and that is the direction that the sentinels watch most carefully.

Ground and aerial predators present very different dangers. Ground predators can be driven off by co-ordinated counterattacks, while raptors can only be avoided (though, if the worst comes to the worst and a mongoose is captured, a rescue attempt can be mounted). The sentinels 'tchrrr' for ground predators and raptors perched in trees, and 'tchee' for flying raptors, which are by far the worst threat.

The threat from predators increases as the group moves faster and forages over a wider area, and also when its attention is focused on a small, concentrated food patch. As long as the group has enough members, extra sentinels will be sent up, perhaps as many as four at a time, during these periods of increased risk.

Nearly all of the sentry duty falls on the young adult males, who spend 20–30 per cent of their potential foraging time watching for danger, and also suffer a 30 per cent higher predation rate. Predators strike just after a sentinel has finished a duty spell, as he runs the 40-60 metres to catch up with the group and his foraging. During that solitary run he is not covered by the new sentinel, who is concentrating his attention on threats to the main group.

The sentinel system is integral to dwarf mongooses' survival, and groups of less than five adults are doomed: because they have to leave gaps in the sentinel schedules they will lose all their young, and one by one their adult members, to

predators. A dwarf mongoose group is a textbook case of social behaviour driven by reciprocal altruism and kin selection (see Chapter 1). Because only the alpha pair breeds, and there is only limited exchange of members between groups, the group is an extended family, and a sentinel, caretaker, or babysitter is caring for its full siblings, which are as closely related to it genetically as its own offspring would be.

Suricates are substantially larger than dwarf mongooses, but they occupy a very similar ecological niche towards the dryer, western half of the subregion. They face the same continual threats from predators, and have an equivalent system of sentinels, but duties are shared more equitably – everyone except pregnant or lactating females takes a turn – and duty spells can be up to an hour long. Unlike dwarf mongooses, suricates frequently swap groups; both males and females leave the group that they were born in and join up with another band, quite often on a neighbouring territory. Because there is such frequent interchange of members, the genetic relatedness within groups is not much higher than in the population as a whole, which rules out kin selection as the major force that drove the evolution of sociality in this species. It appears more likely that suricate sociality is based on reciprocal altruism – by caring for a youngster, even one that is not its close relative, an adult ensures that the group will have enough sentinels, caretakers and babysitters in the future, by which time the caring adult may have become a breeder, producing babies that its erstwhile charges will help to care for.

PORCUPINES

Porcupines live in small colonies with up to about 14 members. Only the alpha pair breeds, but care of the young is shared among all the colony members. When youngsters begin to forage for their own food they are accompanied by one of the colony's adults, probably for defence against predators. Grouping is probably not related to clumped food supplies: porcupines are catholic in their choice of vegetable foods, and they can find food almost anywhere, even if they have to commute up to 12 km to get it. The rare and localised resource that probably does account for their living in groups is the large burrow system that they retire to during the day, which offers shelter from the weather and refuge from predators. In fact, large burrows are so scarce in some areas that warthogs and occasionally even spotted hyaenas share them with the porcupines.

Given their covering of needle-sharp quills, it borders on the bizarre that one of the social glues that bond a porcupine colony is recreational sex. Porcupines are among the few species of mammals that indulge in sex outside the period when the females are reproductively fertile. All the porcupines in a colony mate, but only the alpha female conceives. In the rest of the group, reproduction, but not copulation, is socially suppressed.

MOLE-RATS

If degree of sociality is measured by how profound its effects are on an individual's behaviour and reproduction, the most social mammals are found among the most obscure – mole-rats that spend their entire lives underground. The most social of them all are naked mole-rats from the semi-deserts of Ethiopia, Somalia and Kenya.

In the southern African region the most intense sociality is found in Damara mole-rats. These creatures live in colonies with up to 41 individuals sharing a burrow system. In each colony there is a single breeding pair, and most of the colony members are workers, only just over half the size of the breeders. The workers' reproduction is permanently and irreversibly suppressed; they can never breed in the colony where they were born, and only about one in ten of them ever has even the opportunity to disperse to found a new colony. In all other mammal societies the non-breeding lower ranks have at least some chance that they might one day rise to breeder status, or be able to disperse to another group. Not so for 90 per cent of worker mole-rats.

This extreme form of sociality has evolved in response to the natural selection imposed by the Damara mole-rats' underground habits (themselves probably a way of escaping predation on the surface) and the arid habitats in which they live. Mole-rats of other species living in wetter areas, such as the common mole-rat, also live in colonies that share burrow systems, but they do not have the extremes of reproductive suppression, inbreeding and limited dispersal that are found in Damara mole-rats.

Damara mole-rats live in colonies because they cannot survive alone. They forage by digging, tunnelling through the soil in search of roots, underground stems and tubers. In their arid habitats the soil is as hard as concrete during the long dry seasons, and all tunnelling has to be done in the short periods when the soil is moistened by rain. Even then it is a slow, laborious process that depends critically on teamwork. One mole-rat loosens soil at the head of the tunnel with its incisor teeth, then kicks it backwards to another, who in turn passes it on to yet another until the end of a chain of shovellers, when the last mole-rat pushes it up to the surface. Because their habitat is arid, the underground plant parts on which the mole-rats feed are large and succulent, sometimes weighing several kilograms, and so one lucky strike can feed a whole colony for several weeks. During seasonal bursts of digging that can shift 2,5 tons of soil a month to excavate a kilometre of burrows, a colony can find enough food to supply it for the whole year.

Opposite, top: Banded mongooses live in groups of up to about 30 animals. They do not post sentinels like suricates and dwarf mongooses, but rely instead on the extra vigilance provided by many eyes, ears and noses.

Opposite, bottom: The higher the risk from predators, the more suricate sentinels are usually on guard.

ELEPHANTS

Elephants are one of the very few species in which females live on after their reproductive lives are over. The evolutionary advantage to this post-reproductive survival appears to be the cultural transmission of the store of knowledge that the old cows have built up over their lifetimes. They know where water-holes are that still hold water during droughts, and where food might be available when local supplies are exhausted. The younger members learn from them by demonstration and example, following them and memorising the route as they go.

The importance of culture in elephant behaviour has been demonstrated dramatically by the problems that were experienced with the first attempts to establish new elephant populations by translocations to areas where elephants used to occur but had disappeared. The first of these attempts was made between 1979 and 1983 as part of the restocking of the Pilanesberg National Park with elephants from the Kruger National Park. Because at the time there was no other way to do it, these elephants were cull orphans between four and eight years old – old enough to survive without their mothers, but still small enough to be handled and moved.

When they were released it soon became obvious that their behaviour was abnormal. Among other problems, most of them remained in one small area rather than moving around in search of food and water. At the same time, two ex-circus elephants, Owalla and Durga, were looking for a home, and when they were

An elephant herd approaches water with its youngsters surrounded and so protected by adult cows.

introduced to the herd of youngsters they were remarkably successful as surrogate mother figures for the translocated orphans. They joined up into an approximation of a natural herd, and began exploring over a larger area and sampling the park's vegetation, learning what was edible and what not.

Below and bottom: A baby elephant mired in a mudhole gets a helping pull from its mother.

As the translocated elephants matured some of them became exceptionally aggressive, with frequent attacks on tourist vehicles; but the most bizarre manifestation of the disruption of the young elephants' socialisation was one young male who attached himself to herds of white rhino. This turned from an intriguing aberration into a serious conservation problem when dead rhinos began to be found. By the time the carcasses were discovered it was too late to reconstruct the course of events from spoor, but all the rhinos carried fatal stab wounds in their backs which could only have been caused by an elephant's tusks. A young male elephant was seen pursuing rhinos and apparently trying to mate with them. With the rhino death toll standing at 10 or more per year, more than the average annual loss to poaching in South Africa, the decision was taken to shoot the elephant. Since his removal, rhino killings have stopped. Criticism in the glaring light of hindsight is all too easy, but the elephant populations in Pilanesberg and nearly 40 other protected areas

simply would not have existed today if it had not been for the translocation of cull orphans. In 1996 six mature elephant bulls from the Kruger National Park were released into Pilanesberg in a, so far successful, attempt to socialise the Pilanesberg males and curb their precocious aggression and sexuality. The experiment has been a success; the young bulls' musth periods have shortened and fragmented.

Perhaps just because it is so very conspicuous, elephants' intense sociality tends to be taken for granted, without asking what function it serves. Why are elephants social? They do not feed co-operatively – when cows and calves feast on the camel-thorn pods that a bull shakes loose from the tree it is more exploitation than co-operation – and most elephants live in places where there are no camel-thorn trees for the bulls to shake. They do benefit from one another's efforts at digging wells in river beds, but huge populations of elephants live in areas where there is never a shortage of water, and they have the same social structures as elephants in semi-arid areas. And adult elephants do not suffer the third selective pressure that drives animals of other species to live in groups; the threat of predation – so what is the function of their living in groups?

Top and above: Among other things, the elephant uses its trunk as an important means of communication by touch and smell. The trunk is used for caressing, especially between mother and calf. Elephants can probably recognise each other, assess emotional states and recognize group membership from the smell of their temporal gland secretions.

Cultural transmission of information about the environment is certainly one major advantage of elephants' sociality, and it links sociality with elephants' exceptionally long life spans and their survival long after their reproductive lives are over. Females stay in their family groups, learning from their mothers and older female relatives. Young males are expelled from the breeding herds at 12–14 years old, when their adolescent sexuality and boisterous aggression begin to irritate the mature cows. They join bachelor herds with up to about a dozen bulls of various ages, often including some aged patriarch. These groups are fluid in size and composition, with single animals or splinter groups leaving and rejoining continually. When an old bull associates with two or three younger ones, so-called askaris, it is likely that the askaris learn from their aged mentor some of what the young cows in the breeding herds learn from their matriarch.

Females are far more gregarious than males, and family groups are much more stable than herds of bulls, two differences that provide a clue to the function of sociality. Newborn baby elephants stand 90 cm tall at the shoulder and weigh 120 kg. Their movements are uncoordinated, they are slow, wobbly and painfully vulnerable, and a convenient size for lions or spotted hyaenas to kill. A charge from an elephant cow has no trouble in putting a lion or hyaena to precipitate and undignified flight, but unless the baby can keep up with its mother her charges leave it completely vulnerable to a snap attack by other members of the hyaena clan or lion pride. The predators do not have to kill the calf outright: to wound it sufficiently for it to weaken and begin to lag behind its mother will serve their purposes equally well.

But if there are other adult elephants close by, who can either chase predators or stay with the calf while its mother vents her aggression, the calf is still protected. If there is a threat from predators an elephant breeding herd bunches together with the calves inside a protective phalanx of adults' legs. In Zimbabwe's Hwange National Park there is at least one breeding herd that approaches water-holes with its youngsters protected by two parallel columns of adult cows. Collaborative defence of perilously vulnerable babies is probably what drives elephants up onto one of the peaks of animal sociality.

Only hunters that operate in groups, the super-predators, are able to offer any threat at all to an elephant calf with its mother; even if she has no adult companions to help her, all she needs to do to protect her baby against any solitary predator is to stand her ground. And of the social hunters only the two largest – spotted hyaenas and lions – regularly take prey as large as a baby elephant. The social defensive behaviour by elephants that we see today may be the result of an evolutionary arms race between the world's largest herbivore and Africa's two biggest super-predators. What is more intriguing still is that humankind's early evolutionary history is intricately woven in with the predators and large herbivores that shared our ancestors' African habitats – how much of our sociality is the result of selective pressures exerted by lions and hyaenas? And are clans and prides in any way an evolutionary response to competition from early man?

Under the protection of its mother a baby elephant has nothing to fear.

CHAPTER SIX

COMMUNICATION

signalling and sensing

WITHOUT COMMUNICATION THERE CAN BE NO SOCIAL BEHAVIOUR, and since mammals are so social it is no surprise that they are also intensely communicative; sending messages that are received through all five of the senses.

SIGHT

From our human perspective, communication among other mammals appears to be mostly by visual signalling. During our own daytime activity we see the other mammals that are active in daylight, and to us their signals are obvious and fairly easy to interpret. In fact, with 70 per cent or so of mammal species being small, nocturnal bats and rodents, it is a rather small minority of mammal species in which visual communication has much importance.

For primates like us, active during the day – the monkeys, baboons and apes – communication by visible signals is more important than it is in any other group of mammals. Their facial expressions can signal all degrees of aggression, fear, contentment, submissiveness, curiosity ... in fact, the whole range of what we recognise in ourselves as emotion (which is not to say that we can conclude as a result that monkeys and apes feel the same emotions that we do).

At a more concrete level, the striking contrasts of pattern and colour on monkeys' faces and heads, like the white-rimmed black of a vervet's mask, are unmistakable signals of species identity. A vervet sentinel sitting in a tree sends an all-clear signal just by being there, where the troop can see his bright white undersides. If danger threatens he can give an alarm signal simply by moving away; only if danger appears suddenly and close to his troop will he give an alarm call.

A female baboon's swollen pink behind is a signal of sexual receptiveness that can be seen from hundreds of yards away. The pink colour is critical, and obviously depends for its effectiveness on baboons' having colour vision. She supplements her charms by a seductive slow blink that shows her white eyelids. One of the most striking uses of colour for communication is surely a dominant male vervet monkey's signal of his social rank – a bright red penis set off against the bright blue background of his scrotum. The colour develops at puberty, as testosterone levels rise.

It is no coincidence that squirrels are the most brightly coloured rodents and also among the very few that are active during the day. Their long, bushy tails are particularly important in signalling – all the southern African squirrels combine their chattering predator alarm calls with vigorous up-and-down movements of their tails. Tree squirrels also flick their tails while giving the loud chuck-chuck call that advertises their occupation of a territory.

Large antelope that live in open country typically have striking coat patterns. A gemsbok seen from the side is outlined in black, with a black stripe down its cheeks that runs into and accentuates its horns. A sable bull's uniform black coat makes him look bigger to human eyes, and perhaps to other sable bulls that he hopes to dominate. The white on his face and the bright russet on the backs of his ears are in striking contrast to the black, making the angle at which he holds his head and the

Previous pages: With chewing movements of its bared teeth a zebra signals appeasement and submissiveness.

Below: A female baboon signals her sexual receptivity to every male in her troop.

position of his ears unmistakably obvious. His horns, as well as being lethal weapons, also contribute to visual signalling. A dominant bull stands sideways-on to his opponent, with his tail stuck out stiffly behind, his neck arched, ears cocked, chin tucked in and horns tipped forwards. A subordinate one keeps his tail down, holds his ears back, lowers his head and holds his horns back.

The dark chocolate-brown stripe that separates the cinnamon of a springbok's back from the white of its belly accentuates the stretched and hunched postures that a territorial ram adopts when he urinates and defecates in the middens that mark his territory. During the rut, territorial springbok rams horn the vegetation, sometimes collecting a headdress of greenery. Whether the female springbok that the rams court as they pass through their territories are favourably impressed by these adornments is impossible to tell.

Although they are not as strikingly patterned as gemsbok, sable or springbok, other open-country antelope such as tsessebe, red hartebeest and Lichtenstein's hartebeest also use visual signals. Territorial bulls advertise their status, and their occupation of an area, by standing prominently on raised patches of ground, often old termite hills. Similar behaviour is seen in klipspringer rams, which stand prominently on boulders.

A male vervet monkey's colourful genitals advertise his status.

Some of the antelope that live in denser habitats also send visual signals. Waterbuck and reedbuck have a pale band across the throat that accentuates the position of a male's chin as he stands with head high in the territorial pose.

In all species of spiral-horned antelope – kudu, bushbuck, sitatunga, nyala and eland – there are definite differences between males and females in coat colour. This is most striking in some races of bushbuck where the rams are dark chocolate and the ewes bright russet, and in nyala where the bulls are slate grey and the cows bright chestnut. As they mature, eland and kudu bulls become progressively greyer, especially on the neck.

In an exception which severely proves the rule that it is open-country antelope that make the most use of coat patterns for visual communication, the most colourful and strikingly patterned antelope in southern Africa are nyala bulls, which live in woodland. When a bull is 14 months old he starts to change colour from his chestnut juvenile coat, and by the time he is two years old his flanks are slate grey to dark brown, with up to 14 distinct white stripes across his back and down his flanks, and white spots on his thighs and belly. His long ears have a reddish tinge on the back. The bottom half of each leg is bright yellow with a dark brown or black band separating this from the colour of his body. He has a mane of long, white-tipped hair running from the top of his head to the root of his tail, and the white tips to the hairs produce a white stripe down his back when the mane lies flat. From below his lower jaw, down the underside of his neck, along his belly and onto the undersides of his thighs there is a fringe of long dark hair.

He parades his colours in a visual dominance display. At low intensity he slightly raises the crest of long hair along his back, which increases his apparent size and produces a sharp contrast between his dark flanks and the white tips of the mane. At high intensity he fully erects his mane, holds his head high and parades slowly with high steps of his yellow legs. In the full display he also curls his tail up over his back and fans out the white hairs on its underside, and holds his head low with his horns pointing forward. A submissive male lowers his crest, waves his head from side to side and slowly wags his tail. He will break off an encounter by grooming or feeding.

Fluffing up the hair, and raising manes and crests are common threat signals for either defence or offence. Their main effect is to present a more formidable appearance to deter an opponent or attacker. An aardwolf, white-tailed mongoose or large grey mongoose can increase its apparent size by about a third by raising the long hair on its back and tail. When a porcupine raises its hair it not only increases its size but also displays its armoury of defensive spines.

Baring the teeth is an almost universal signal of aggression. Primates, carnivores, rodents, zebras, some whales, insectivores, elephant shrews, dassies and hippos all use their teeth as weapons, and expose them in both defensive and offensive threat.

Opposite, top: Nyala bulls parade their colours in a dominance display that will escalate to fighting if neither backs down.

Opposite, bottom left: A gemsbok's coat patterns emphasise its outline and head position; the closer of the two shows the more extreme dominance pose.

Opposite, bottom right: The solid black of a sable bull's coat emphasises his size.

Top: A hippo harem master threatens an intruder to his territory by displaying his tusks, including a pair of lower canines that can reach lengths of 30–50 cm.

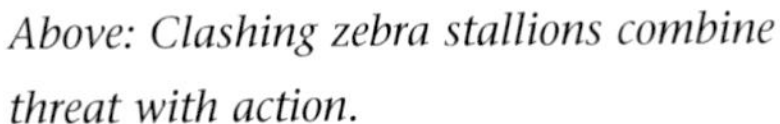

Above: Clashing zebra stallions combine threat with action.

Above, right: An elephant bull in his prime displays his weapons.

The most spectacular of these displays is a hippo bull's threat yawn, in which he opens his enormous mouth to display the battery of tusks in his upper and lower jaws, which includes a pair of massive lower canines whose tips are kept sharp by wearing against the canines in the upper jaw.

The most impressive dental weaponry – the pair of tusks carried by African bull elephants – is kept on permanent display. With maximum lengths of three metres, they are used by bulls as weapons of offence in the rare but spectacular battles between two musth bulls (*see* Chapter 7) and by both sexes as weapons of defence against predators, including man. Females and young males jab each other with their tusks in the skirmishes around water-holes. Both bulls and cows hone the tips of their tusks to lethal sharpness by rubbing them on trees, and bulls in their prime between 25 and 40 years old who are regularly going into musth have much sharper tusks than older bulls who are past their reproductive peak.

Paradoxically, baring the teeth can also be a signal of submissiveness. A submissive Burchell's zebra makes chewing movements with its teeth exposed, and a primate's fear grin exposes the incisors but not the canines. Jackals bare their teeth in both aggressive and defensive threat – the interpretation of the message rests in the position of its ears, pricked forward in aggression and laid back in defence.

Ratels, white-naped weasels and striped polecats all have well-developed anal glands whose potently odorous secretions they use in defence. They advertise this chemical weaponry by having coats strikingly marked in black and white. The most adept of the three at chemical warfare is the striped polecat, which can squirt the secretion of its anal glands in an aimed stream that can temporarily disable an attacker by choking and blinding it. If a polecat's attempts to escape are thwarted, it arches its back, lifts its tail, and fluffs up its long, silky fur as a final warning that it will deploy its chemical arsenal if it is provoked any further.

In what might be a visual bluff, for their first three months cheetah cubs have dark grey undersides and mantles of long, pale grey hair, which may make them look enough like foul-tempered ratels to deter some potential attackers. Nonetheless, 70 per cent of cheetah cubs in the Serengeti are killed by lions.

A zebra's stripes are not only a means of making it more visible to its herd companions (*see* Chapter 4); they also act as a focus for the mutual grooming that is an important group bonding mechanism. The most popular grooming sites are the legs, shoulders and neck, where the stripes narrow. A mountain zebra male rests his chin on the gridiron stripe pattern on his companion's rump.

A kudu bull's mane accentuates the thickness of his neck, which is itself the best indicator of his fighting ability. Mature eland bulls develop dewlaps that further emphasise their massive necks and shoulders. A lion's mane gives him an even more formidable appearance, as well as protecting him during fights.

Hares and antelope of several species have bright white undersides to their tails, which they display as they flee from predators. These white flashes may be an alarm signal to other prey, whose simultaneous flight might confuse a predator, and they may be a way of making it easy for other fleeing prey to follow close behind, to the benefit both of the leader (which now has another animal between it and the pursuing predator) and the follower (with an animal between it and a predator that might be lying in ambush). Warthogs flee with their tails stiffly upright, probably so that they can follow each other in long grass. White flashes may also be a signal to the predator that it has been seen, and has lost the element of surprise and with it its chances of making a kill. By providing an easily visible target it may even tempt a predator into attacking too soon or from too far away. Communication with the predator is the most likely function for the white tail scuts of fleeing hares: they do not live in groups and so have neither use for alarm signals nor opportunity to signal other hares to scatter or to follow behind. Visual communication with predators is also the function of stotting and pronking, the exaggerated bouncy gaits that some prey animals use when they are chased by a predator, or before a chase begins (*see* Chapter 4).

Below: White tail flashes may act as alarm or follow-me signals.

Top: Contrasting patterns make facial expressions more visible in poor light.

Above: The position of a wild dog's tail, accentuated by the white tip, sends social signals. Arched up and forward as here, it signals excitement and aggression.

Leopards have a white patch on the underside of the end of their tails, which is believed to make it easier for cubs to follow their mothers after dark. Servals, whose tails are much shorter than a leopard's, have black and white markings on the backs of their ears instead, but these are more likely to have the function of making the ears' positions more obvious during social encounters in the gloom of the thick cover which is servals' favoured habitat. The white patches that genets wear on their faces, a civet's black bandit's mask and the black backs and long tassels of a caracal's ears are there for the same reason.

Like servals, lionesses also have spots on the backs of their ears, which may help them to co-ordinate their hunting manœuvres by making them more visible to one another. Wild dogs' white tail tips may have an equivalent function during their long chases, and they also accentuate the tail's position and movements during social interactions.

SOUND

As a channel for communication, sound's essential properties are that it can change very quickly and be turned on and off abruptly; it carries messages over considerable distances even in broken country or dense cover; it can broadcast messages in all directions, and it can be detected even when the receiver does not have its attention on the sender – it is possible to hear without listening. A lion's roar can be heard even by a human at a range of 8 km; elephants respond to one another's infrasound calls over distances of at least 5 km. Sound carries even better underwater than it does in air, and the calls of large whales like humpbacks can travel for thousands of kilometres.

Sound's particular properties make it the best of all communication channels for sending alarm signals. Rabbits, elephant shrews and black wildebeest stamp their feet, and antelope of nearly all species use an explosive snort through the nostrils as an alarm signal, often repeated several times while the animal watches a potential predator. A klipspringer's alarm signal is a piercing whistle, which signals to its mate that there is danger close by, but also tells the predator that it has been sighted. Bushbuck and kudu give alarm barks with a bewilderingly ventriloqual quality that makes their source almost impossible to locate. Interestingly, male antelope also give alarm snorts during aggressive encounters. While these may truly reflect unease, there is a possibility that their function is to unnerve the opponent, with an equivalent of: 'Look out for that lion!'

For animals that are the prey of others, using sound for communication has a major drawback: it very easily attracts the attention of predators. Therefore, as a general rule, carnivores are more vocal than herbivores, and signalling by sound is more important in those herbivores that are well protected against predation – elephants by their size, hippos in water, and arboreal squirrels by their alertness and agility in the trees. In an apparent paradox, the most vocal of the carnivores are suricates and dwarf mongooses, whose small size and daytime activity make them very vulnerable to predators. The paradox is resolved by their continual contact and all-clear calls being essential components of the sentinel system that protects foraging groups from predation (*see* Chapter 5). The exception that proves the rule about prey being less vocal than their predators is provided by territorial impala rams during their rutting season (*see* Chapter 7); they can be heard snorting and roaring day and night, and they also suffer heavier predation – the sex ratio among adults is two females to one male.

Elephants are extremely vocal. Because they have little reason to conceal themselves from predators they can afford to keep in touch by long-range contact calls and to communicate within herds by a range of vocalisations that are graded in intensity and length. At crowded water-holes there is a continuous cacophony of rumbles, growls, trumpeting and squeals. Loud growling and roaring are threats; bellowing signals sudden pain or fear, as when one elephant jabs another with its tusks. Young elephants squeal in distress. Trumpeting is used in a variety of contexts – excitement, anger, alarm and greeting – and how other elephants interpret it almost certainly depends on subtle differences in sound and other signals, such as ear position, that the trumpeter sends at the same time.

Even casual observation of elephants shows that there are aspects of elephant vocal communication that humans are deaf to, both figuratively and literally. Most of elephant vocal communication uses infrasound, with frequencies below the lower limit of human hearing. The growling 'tummy rumbles' that are so characteristic of elephants are only the upper harmonics of infrasonic contact calls. The calls are very loud, around 110 decibels, which is about the same as a lion's roar. Where habituated elephants allow close approaches by vehicles it is sometimes possible to feel the air pulsing as a nearby elephant gives a loud infrasonic call. Low-frequency sound is not strongly attenuated by air or scattered by vegetation and elephant calls can certainly carry for at least 5 km. The more extravagant claims for ranges of 30 km or more are not backed up by reliable evidence.

Bulls in musth (*see* Chapter 7) are especially vocal, though at frequencies largely inaudible to humans – the part of one call that we can hear sounds like water sloshing around in a long pipe. Cows are strongly attracted by a musth bull's vocalisations, and bulls that are not in musth are repelled by them.

At the other end of the size scale are the shrews, insectivores, bats and small rodents that communicate by ultrasound. Because ultrasound attenuates very quickly in air their calls carry only short distances, but an animal that is itself only a few tens of millimetres long has no need to broadcast its messages over vast areas.

Brants' whistling rat, found in the Karoo and Kalahari, gives a piercing whistling call when sensing danger.

Although the sounds that insectivorous bats use for echolocation have to have frequencies well above the range that humans can hear (*see* Chapter 3), some of their social calls have much lower frequencies, giving them more carrying power, and these are audible to humans. Colonies of Commerson's leaf-nosed bats are particularly noisy as they squabble for resting spaces after a night's foraging, but the most familiar bat call is probably the repeated, musical 'tink' given by male epauletted fruit bats while hanging in trees. The function of the call is not certain, but it probably has something to do with maintaining the spacing between males, who do not call from less than 200 m apart.

Long-range calls have two contrasting functions, to draw animals together, and to keep them apart. The most familiar recruitment calls are those of the social carnivores: a lion's roar, a spotted hyaena's 'whoo-oop', a wild dog's musical and resonant 'hooo', and a black-backed jackal's drawn out 'nyaa-a-a-a' all serve to establish contact between group members, and to allow them to locate and meet up with one another. Attractant calls are used in some species to attract mating partners. Leopard males give a rasping cough that sounds exactly as if a thin plank is being cut with a coarse saw: its function is both to attract females, or at least to alert them to the male's presence, and to warn other males off from trespassing in the caller's territory. Spacing calls, which serve to advertise an animal's presence and location so that others can avoid it, are often but not always associated with territoriality. The roars of pride male lions, the sawing calls of territorial male leopards, the deep 'boom' from a male samango monkey, and the grunts of hippo harem masters serve to warn other males that an area, and the females within it, are spoken for. Elephant bulls that are not in musth avoid those that are by listening for their infrasonic rumble calls.

Above: Male epauletted fruit bats give repeated musical 'tink' calls at night.

Below: The roars of male lions are individually distinct, and the roars of a lion calling to establish contact with his pride may be different to those that proclaim his territorial status to other males.

In some cases, what sounds to human ears like a single type of call may stimulate clearly different responses in animals that hear it. Vervets are capable of very subtle discriminations between calls: harsh grunts that to human ears all sound the same are used in four different contexts; by a subordinate approaching a dominant, a dominant approaching a subordinate, as the troop moves away from cover, and when another troop is sighted. Monkeys that hear the calls, or tape-recordings of them, respond differently to each one. The roars of male lions are individually distinct, and given the hidden complexities that have been discovered in the calls of other species it would be no surprise to find that the roars of a lion calling to establish contact with his pride were different to those that proclaim his territorial status to other males.

Among cats the ability to give loud, low-pitched calls like a lion's roar or a leopard's sawing call depends on the larynx being connected to the skull only loosely by a strip of cartilage. In the small cats the larynx is anchored by a chain of small bones that limits them to quieter, higher-pitched sounds.

Galagos or bushbabies, which are small and active at night in dense cover, have complex repertoires of calls that can be graded and combined, and galago species are easier to recognise by sound than by appearance. Southern Africa has three species: the thick-tailed bushbaby, largest of the three, has a call that sounds uncannily like a wailing human child, repeated 3-17 times at regular intervals. A lesser bushbaby gives a long series of barks lasting up to an hour, quite different from the call of Grant's lesser bushbaby, which is very similar in appearance to the lesser, but which gives a sequence of up to 18 double or triple calls lasting 3-6 seconds, reaching a crescendo over the first few calls and then trailing into a series of deeper, rapid staccato notes. The dramatic differences between the calls leave other bushbabies in no doubt over whether it is an animal of their own species that is calling.

Stamping the back feet is an alarm signal for rabbits and hares, a contact signal for elephant shrews of most species, and an alarm call for bushveld and rock elephant shrews. That animals as dissimilar as rabbits and elephant shrews should use such similar signals appears puzzling but their behaviour actually fits in with modern genetic evidence that the elephant shrews' closest living relatives are indeed rabbits and hares.

Stamping the back feet is an alarm signal used by rabbits and hares. The scrub hare (above) also squeals when distressed.

Sound plays a large part in threat behaviour, both offensive and defensive. Fights between members of the same species – over food, resting sites, females, social status or whatever – are nearly always accompanied by a tumult of noise. Fighting rhino bulls snarl, grunt, growl and scream. Warthog boars grunt and squeal when their territorial disputes escalate to attempts to slash one another with their lower tusks. Any dispute in a troop of baboons or monkeys is accompanied by a cacophony of screeching. The trumpets and squeals of elephants at a water-hole that cannot supply them all, or the tumult of roars and snarls around a lion or spotted hyaena kill, are mixtures of vocal threats, alarms and reactions to sudden fear and pain.

Noisy defensive threats are used in some species for defence against predators. A porcupine under threat roars, stamps its feet, rattles its quills and makes a dry, hissing rattle by shaking the tuft of special hollow quills on the end of its tail. A striped polecat's threat of the imminent use of its anal gland spray combines hair fluffing and tail arching with a stream of high-pitched screeches. In some cases the threats are only bluff: an aardwolf's loud roar, combined with its raising of its dorsal crest, gives an intimidating impression of a large and ferocious creature, but although it has a respectable set of canines, the rest of its teeth are small and ineffective.

Although they feed on land, hippos spend at least two-thirds of their lives in water, and nearly all their social interactions take place there. Harem bulls proclaim their status with a distinctive repeated snort, given as they surface. Neighbours echo the call, and waves of snorting pass from one pod to the next. The status call also travels under water, certainly faster and probably further than it does in air.

While they are submerged, hippos of both sexes make a variety of clicks, croaks and whines that cannot be heard above the surface. Some of these noises might be used for echolocation, because the water that hippos live in is very often thick with mud and there is no way that they could find their way around underwater by sight.

For whales and dolphins, sound is almost certainly the most important channel of communication. Southern right whales bellow loudly when at the surface, and 'belch' and 'moan' when submerged. Dolphins communicate by clicks and whistles. Sounds are not produced only by vocalisations: southern right whales use breaching, flipper smacking and lobtailing, in which the tail flukes are slapped down onto the surface, to generate underwater sounds. What messages the calls and other sounds send, and how they fit into social structures, is a long way from being understood.

Vervet monkeys give different alarm calls for different sorts of predator, and the responses of monkeys that hear the calls are adapted to the sorts of danger that each predator presents. A short, sharp chirp that is given by youngsters and females warns of a dangerous mammalian ground predator such as a leopard, and the response is to flee into trees and climb out onto branches that are too thin to support a leopard's weight. 'Praup' signals a sighting of a large bird of prey, which is a threat to monkeys on the ground in the open, who flee into thickets, and those on the outer branches of trees, who climb in towards the centre. At night, large owls are harried with two barks and a growl, repeated over and over. 'Uh' is a low-intensity alarm given for predators like spotted hyaenas and wild dogs that pose no serious danger. Alerted monkeys keep watch on the predator or move slowly away. 'Nyow' is a startled response to any sudden nearby movement. The snake alarm is a 'chutter' which makes other monkeys stand up and look around – they are especially wary of poisonous species and pythons. In areas where vervets are afraid of humans they chutter for them as well, but the call is slightly longer and lower in pitch than that given for snakes.

Of all the sounds produced by southern African mammals, the most puzzling is the clicking noise that eland bulls make while they walk – its source is uncertain, and its function, if it has one, is a complete mystery.

A waterbuck bull assesses a female's reproductive condition from the smell of her urine.

SCENT AND TASTE

By far the most important channel of communication between mammals is through chemical signals: blends of dozens or hundreds of organic compounds that are perceived as odours or tastes, or by specialised sense organs. Of all the ways by which mammals send messages, chemical communication is the least understood and the most difficult to study, mainly because our human sense of smell is so miserably inadequate in comparison with the scenting abilities of almost all other species of mammals in the animal kingdom. We do not even have the language to talk sensibly about smells.

Airborne organic volatiles transmitted from one mammal to another influence dominance hierarchies, spatial organisation, foraging patterns, mate choice, sexual maturation, mother-infant interactions, individual recognition and food choice. Chemicals can signal the sex, age, social status, group membership, reproductive condition, kinship, individual identity, diet, species or subspecies and emotional state of the animal that produced an odour. Reproductive suppression in social groups of species such as dwarf mongooses, porcupines and Damara mole-rats is maintained at least partly by chemical signals produced by the dominant breeding female.

Scent signals may be general body odours, like the goaty smell of a waterbuck or sable bull, or may be produced by special glands. Urine and faeces also carry scent signals: urine especially contains metabolic by-products that reflect diet and hormone levels. Nearly all carnivores have a pair of anal glands where bacteria process skin secretions into smelly brews that usually contain fatty acids and sulphur compounds. Most antelope have preorbital glands that open just in front of their eyes and whose secretions, which are not always particularly strongly scented to the human nose, are deposited on vegetation. Glands on the feet and legs, which mark the soil and vegetation as the animal walks, are common in all sorts of mammals. Rodents have scent glands near their eyes and in the anogenital region, and urine is a particularly important odour carrier in these little animals. Shrews have flank glands whose secretion rubs off onto the sides of their runways. The allogrooming, mutual scent marking and sleeping together that are such important components of social behaviour all serve to build up a group odour that serves as a badge of membership.

Mammalian chemical signals are enormously complex blends of chemicals, and in most cases the messages that they carry are encoded by the relative concentrations of a number of the components rather than by the simple presence or

Above: Many antelope, like this Damara dik-dik, have preorbital glands that open just in front of their eyes. Their secretions are used as territorial scent marks.

Below: Springbok ewes check a lamb's individual odour.

Cheetahs demarcate their home ranges with scent marks of sprayed urine.

absence of single chemicals. The complexity of the signals gives them the potential of carrying huge amounts of very specific information. Just 19 compounds, each with only three different concentrations, could provide a unique odour label for over a billion individual animals. So it is no surprise that, in every mammal species so far tested, including even humans, individual odours can be discriminated and recognised. In the case of the small-spotted genet a female can remember the odour of a male that mated with her nine weeks before.

The spotted hyaena greeting ceremony allows mutual recognition by odour, and like much else about them, it is unique among African mammals and slightly bizarre. The ceremony is always initiated by the lower-ranking of the two participants, and begins with mutual sniffing of mouths, heads and necks. The two hyaenas then move to stand head to tail, and each cocks one hind leg, rather like a dog urinating against a lamp post. This exposes its genitals with erect penis or clitoris, which are carefully sniffed by the partner. The greeting ceremony usually lasts about half a minute, and – like scent-marking – it first appears in hyaena cubs less than a month old, while they are still in their dark grey baby coats.

Scent marks are endowed with one property that no other form of communication possesses: because their messages persist through time they send information into the future. A lion's roar proclaims his status to anyone within an 8 km radius, but only as long as he is roaring. The equivalent message in the form of pungent, tomcat-smelling urine sprayed onto a bush persists for days or weeks. This unique ability to send messages into the future is especially important when animals deploy scent signals in relation to territoriality and their use of space. Dwarf mongooses are fiercely territorial, and any dispute between neighbours is accompanied by a frenzy of scent marking. When the mongoose group moves on from the disputed area its scent marks remain as a token of its claim over the territory. A group moves to a new foraging area nearly every day, exploiting them in rotation so that food resources do not become too seriously depleted. The time that it takes for a group to make one circuit of its territory coincides to within a few days with the time that it takes for its scent marks to lose their odour. It could easily be that the group uses the dwindling odour from the scent marks that it deposited the last time it used an area to regulate how soon it will use it again. Cheetahs can have overlapping home ranges, but they rarely meet because they move away from areas where they find urine scent marks less than two days old. Wild dog packs scent-mark their huge home ranges with urine and evil-smelling droppings. Their notoriously strong body odours probably help lost dogs track their packs.

Scent marks may be deposited in the form of urine, faeces or special scent secretions, and many scent-marking behaviours have been designed by natural selection to put the scent where it is most noticeable. Jackals deposit their faeces on top of grass tussocks, rocks, or even elephant cannonballs. A cat or rhino's backward squirt of urine is targeted at nose level. Damara dik-diks build dung middens along pathways and at junctions. An oribi ram deposits small beads of black preorbital gland secretion on the tops of tall grass stems, and if a stem is too long he bites it down to a suitable height. A steenbok also prunes its marking sites to a convenient height and shape, but it marks them with the secretion from a gland under its chin that is not found in other antelope. A territorial suni ram puts most of his preorbital gland marks on twigs and stems about 6 cm tall, in areas where an intruder would be likely to feed with its nose near the ground.

The prominent, jet black tufts low down on an impala's hind legs are rings of long, glossy, black hair surrounding a patch of thickened, hairless skin, which produces a secretion with a sharp cheesy odour. Despite impala being so common and so easy to observe, the function of this scent gland is a mystery: impala do not sniff at the glands on one another's legs, and they do not show any behaviour that might transfer the scent to vegetation or the soil to produce scent marks. It has been suggested that the scent is released when a herd scatters during a predator's attack, making an airborne trail for other impala to follow – a sort of olfactory equivalent to a white rump flash – but such a trail would be too easily erased by the slightest of breezes for it to be of any use.

An impala's ankle tufts surround glandular patches, but their function is a mystery.

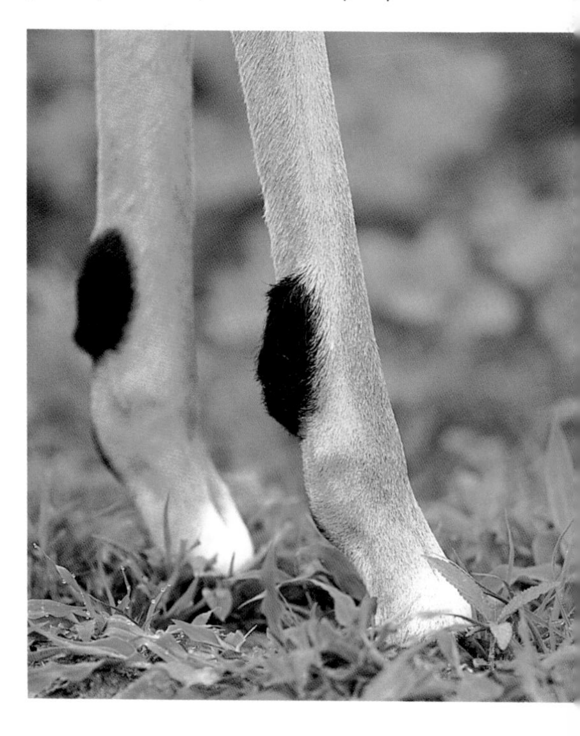

Brown hyaenas use an elaborate marking behaviour known as pasting to produce unique scent marks with two distinct parts. The hyaena walks over a grass stem, which bends down under its belly, and when the stem is between its hind legs the hyaena crouches and extrudes the fist-sized anal gland that opens just under its tail. It manoeuvres a groove in the gland down around the grass stem and then withdraws the gland as it steps forward, leaving the grass stem marked with a smear of dark secretion and, about a centimetre below it, a blob of creamy, white paste.

The two pastes appear to have two distinct functions – as part of the hyaenas' nightly foraging for carrion, and as signs that a specific area is part of an occupied territory. The black paste loses its odour fairly quickly: after about 7–10 days it has no smell to a human nose, and its function appears to be to remind the hyaena that deposited the mark, or to inform its group companions, that the area has been searched for carrion within the past few days, and so richer pickings might probably be available elsewhere.

The white paste lasts much longer than the black: it still has an odour to the human nose after about 4–6 weeks, and its function is probably to inform itinerant hyaenas that the area they are moving through is an occupied territory. On average, a brown hyaena pastes once every 400 metres of the 30 km that it covers in its

Above: A spotted hyaena checks one of its group's scent marks. A ring of scent marks and middens of bright white droppings (below) usually mark each hyaena clan's territory boundary.

nightly search for carrion. A territory in the Kalahari region may be peppered with 20 000 active marks, and so a hyaena would have to walk only about 500 metres before it encountered one.

Although their anal glands are smaller than a brown hyaena's, spotted hyaenas use a roughly similar behaviour to leave smears of brown secretion on several grass stalks at once. These marks are deposited along territory boundaries, very often near middens of the hyaenas' prominent white faeces. The middens themselves are boundary markers, and are added to when the resident hyaena clan patrols its borders: the largest middens contain hundreds of faecal pellets, whitened by the bone that they contain.

The hyaenas' smaller relatives, aardwolves, use a similar pasting behaviour to deposit single-component scent marks on grass stems in order to demarcate their territories. The odour of the marks is astonishingly persistent: a mark that is sheltered from the weather still has an odour after about nine months. An aardwolf marks an average of 10–20 times per kilometre as it covers 8–12 km each night in its search for harvester termites. A pair of aardwolves will deposit approximately 300 marks a night, concentrating them near the territory boundary to set up a boundary line, and each aardwolf territory can be showered with hundreds of thousands of active scent marks.

The odour chemicals in a hyaena's or any other carnivore's anal gland secretions are not produced directly by the animal itself, but by colonies of bacteria that live in the gland and ferment skin secretions into potent brews of fatty acids, sulphur compounds and amines, all with very potent odours.

Middens are accumulations of droppings, very often mixed with urine. Their occurrence is sporadic – rock rabbits use them but hares do not, some mongooses and small cats use them, big cats and canids do not. Middens are very common among territorial antelope, and hippo and rhino bulls deposit prominent dung piles, but giraffes, elephants, zebras, warthogs and bushpigs scatter their droppings wherever they happen to be. The first thing that an aardwolf does when it wakes in the evening is to make a beeline for a nearby midden where it deposits and buries a large pile of faeces that can weigh 10 per cent as much as the aardwolf itself. The faeces consist mainly of sand and termite heads, and they smell strongly of turpentine from the termites' defensive secretions. It could be that the aardwolf buries them in one place so that their smell does not distract it while it is foraging.

Rather than accumulating large dung piles as most territorial antelope do, a steenbok mixes individual piles of its droppings with soil by scratching vigorously with its forefeet. Whether partly burying the faeces hides their smell from predators or preserves the smell as a territorial marker is far from certain.

Large and prominent white rhino middens are produced and used by territorial adult males, and usually lie on the border between two territories. When he visits one of his middens a territorial bull sniffs around carefully, stirs up the midden with his hind feet, defecates and then kicks the dung around, gouging up the soil in the process, and producing an area up to five metres across of ploughed-up soil mixed with dung. He also sprays nearby bushes with backward squirts of urine. Females and bulls that do not hold territories may defecate in middens, but they neither spray urine nor kick dung.

Rhino middens serve as information centres, keeping neighbours up to date on one another's presence, physical condition and social status. Despite having different spatial organisations (white rhinos are territorial, black rhinos are sometimes not), both species make middens, and, puzzlingly, often use each other's, without any other sign of inter-specific territoriality. The two species produce similar-sized dung balls, but the white rhino's contain only grass while the black rhino's contain sections of twigs snipped off at a 45° angle.

The largest middens are produced by one of the smaller mammals. By urinating and defecating repeatedly in the same place over several generations, a rock dassie colony's inhabitants build up huge deposits that are sometimes over a metre deep, besides which a white rhino midden, produced by an animal over 500 times as heavy as a dassie, pales into insignificance.

Bushbabies (galagos) have a special scent-marking behaviour called urine washing – the bushbaby urinates over one hand and then rubs it against the

Rhino dung middens act as both territory markers and information centres for neighbouring bulls.

A buffalo bull determines a female's reproductive condition by sniffing and flehmen of her urine.

opposite foot, and repeats the procedure with the hand and foot on the other side. Then, as it climbs among the branches, it leaves a trail of damp, smelly foot- and hand prints. They also mark branches with secretion from a gland on the chest. The gland is larger and more active in adult males, and the scent marks are probably related to territoriality.

Outside the primates, the land mammals that use scent signalling the least are probably giraffes. Bulls assess cows' reproductive status by sniffing and flehmen of their urine, and odour probably plays a role in bonding and recognition between mother and calf, but a giraffe has no discrete scent glands; it does not make scent marks, and it drops its dung anywhere rather than in middens. This very unusual lack of olfactory communication is very likely to be connected to the giraffe's height: its nose is simply too high above the ground to detect and decipher the odour signals from ground level.

With the possible exception of the marine mammals, all female mammals communicate their reproductive status through chemical signals. In particular, their faeces and urine contain the breakdown products of the steroid hormones whose fluctuations drive reproductive cycles, providing a direct indicator of their sexual status. Mammals of many species have a special sense organ to detect and decipher these steroid residues – the vomeronasal organ, also known as Jacobsen's organ, which lies just above the roof of the mouth and is connected to the mouth or nose, or both of them, by narrow ducts. Humans have only a vestigial vomeronasal organ, a deficiency that we share with other higher primates and some bats. The vomeronasal organ is brought into use by the behaviour known as flehmen (a German word with no direct English equivalent), in which the animal curls its upper lip, wrinkles its nostrils closed and moves its tongue back and forward to pump the steroids into the vomeronasal organ.

Flehmen often follows sniffing and licking of a female's genitals and urine by a male with an interest in her reproductive condition. Cats usually touch the tip of the tongue to the urine and then transfer a tiny sample to the openings of the vomeronasal duct. Giraffe and buffalo bulls, and some antelope males, catch urine on their tongues while the female is urinating. Usually the vomeronasal organ is brought into use only after more conventional airborne odour signals have shown that there may be steroids that are worth investigating. In some cases the female animal's willingness to produce a urine sample for testing is itself a sign of impending sexual receptivity.

TOUCH

Communication by touch in its usual sense can operate only over the shortest of ranges; between animals that are in physical contact, but as a channel for long-range communication touch begins to overlap with sound when mole-rats communicate by drumming their feet against the sides of the burrow and detect the vibrations through touch-sensitive hairs rather than by hearing.

Touch as communication also has a fuzzy boundary with direct physical manipulation. When a female cat picks up a kitten by the scruff of its neck there is more to it than the simple mechanical act of carrying, because a reflex is stimulated that makes the kitten hang limp and passive. Is one jackal's hip slam on another an attack, or a tactile aggressive signal?

Because a mammal mother's care of her young always involves a high degree of intimate physical contact, the mother-offspring relationship is one in which tactile communication is especially important. The stimulus of sucking causes a nursing female to let down her milk, and is necessary to maintain lactation; once a litter of kittens have established their ownership of particular teats, the rest of the teats stop yielding milk. Suckling antelope lambs and calves stimulate let-down of their mother's milk by nudging and butting her udder. In baby rodents, carnivores and hider antelope (*see* Chapter 4) urination and defecation are stimulated by the mother's licking so that she can swallow the wastes to keep the nest or hiding place clean and free of the odours that might attract predators. Canid pups nudge and lick the lips of the adults that care for them to stimulate their regurgitation of a meal of meat. This infantile begging forms the basis of greetings between adults, and the food-sharing in a pack of wild dogs.

Touch is a universal element of the bond between offspring and mother.

Above: A leopard cub receives a flannelling from its mother's tongue.

Below: A steenbok ram lifts one front leg to tap the hind legs of the female he is courting.

In social groups the tactile interactions of play, allogrooming, huddling and sleeping together act as social glue by providing immediate tangible rewards for group membership. That social mammals enjoy close physical contact and being groomed is an evolutionary design whose function is the formation of groups – the animals find immediately rewarding what best serves their long-term genetic interests. Tactile interactions also send messages: in nearly all groups, grooming is initiated by subordinates and received by dominants. Alpha female breeding status in a dwarf mongoose group is decided by a grooming contest that may last days, and which ends with both contestants slathered in saliva.

Touch is an indispensable component of courtship in all mammals. Antelope rams and bulls lick and nuzzle their females' genitals; a steenbok ram taps the inside of his mate's hind legs with one of his forelegs; a waterbuck or sitatunga bull rubs his head against the female and presses his chin down on her back as a prelude to mounting. Male nyala butt females between their hind legs, sometimes so vigorously that their hindquarters

are lifted off the ground. Bushbaby pairs first establish social bonds by a frenzy of mutual grooming that can go on for hours at a time. Among viverrids (civets, genets and mongooses), mustelids (otters, striped polecat, white-naped weasel and honey badger) and cats, mating includes a bite by the male to the nape of the female's neck. In lions and some of the mongooses the bite is ritualised, but in mustelids the males use it to pacify and physically control the female before and during copulation.

Above and below: Allogrooming and being groomed are considered immediate tangible rewards for group membership and social co-operation. In nearly all groups, grooming is initiated by subordinates and received by dominants.

As her mate's interest in copulation begins to wane a lioness will start to take the initiative, rubbing sinuously along the male lion's flanks and across his chest. As they mate, touch receptors in the female's genitals send messages to her brain, which in turn triggers the hormonal changes that result in ova being released for fertilisation. To ensure that the message is received and 'understood' the male's penis is armed with tiny, backward-pointing spines. A female cat's aggressive reaction to the male's withdrawal after mating is a reaction to the intense and presumably painful stimulus that the spines produce.

CHAPTER SEVEN

REPRODUCTION

closing the circle

THERE IS A FUNDAMENTAL, INESCAPABLE ASYMMETRY in mammal reproduction: it is only the females that gestate and lactate. Because gestation takes place inside the female's body, fertilisation also has to be internal, and once sperm and egg have fused the female is left, quite literally, holding the baby.

Female mammals invest much more time and energy than males in the development and growth of their offspring. Males invest much more than females in securing opportunities to mate. Typically, a male mammal makes his investment in genetic posterity before the moment of conception, a female makes hers after it.

In the vast majority of mammals, the males' contribution to their offspring is no more than an ejaculation of semen; the females cope alone with the task of nourishing the foetus, giving birth and bringing up the infant. Gestation and lactation are such effective ways of caring for the young that most females are more than equal to the task, and as a result only about 10 per cent of mammal genera have males that help to care for their offspring, and in only about five per cent of mammal species are the males monogamous. What the members of this small minority have in common is that the efforts of both parents are needed to successfully raise the young to independence. This may mean that the male brings food to the female and the youngsters, as male black-backed jackals do; that he helps to protect them directly, as male aardwolves and porcupines do; or that he protects the family's resources, like the rams of small antelope that live in pair territories – suni, klipspringer, red, blue and grey duiker, and Damara dik-dik. If a male left the female to her own devices after mating, their young would not survive and the male would leave no genetic legacy.

Monogamy does not necessarily limit group size to two. Wild dogs, porcupines, dwarf mongooses, suricates and two species of mole-rats live in social groups in which, usually, only one pair breeds. In all these social groups the role of the father as a helper is shared with, or even taken over by, the other members of the group, who are usually the older siblings of the current crop of youngsters.

True polyandry, in which a single female has multiple mates who care only for her offspring, is unknown among mammals. In social groups with helpers, several males may contribute to bringing up the young, but they do so without having mated with the group's breeding female. In brown hyaenas, male group members help raise cubs that were fathered by an itinerant male who is not even a group member. The genetic payoff for the helpers is their close relatedness to the youngsters, who are usually their half-siblings, nephews and nieces.

Previous pages: While she licks it dry a blue wildebeest cow imprints on her newborn calf's odour.

Top: A giraffe cow on heat has a high-ranking bull as a consort.

Above: Mating between springbok is a fleeting affair.

The function of sex is reproduction – with the exception of porcupines, which (somewhat bizarrely) use sex as a social bonding mechanism, and lionesses, which use it to protect their cubs from infanticide (*see* Chapter 5) – females do not mate except when they are in oestrus, and this has a direct influence on the sexual activities of the males.

It is very common, perhaps universal, for a female to become sexually attractive for at least a short while before she is sexually receptive. In many species she will begin to attract the interest of males for some time before she is willing to mate with any of them. A giraffe cow becomes attractive at least 24 hours before she is ready to mate, and she is courted by a series of increasingly higher-ranking bulls as each suitor is displaced, usually with no more than a standing tall display, by one higher up the local hierarchy. By the time that she is willing to mate she will have as her consort the area's highest-ranking bull. Dominant bulls are so effective at finding females on heat and displacing lower-ranking consorts that they father nearly all of the calves.

White rhino bulls also consort with cows that are showing signs of impending oestrus. As cows pass through his territory the resident bull checks their reproductive condition by sniffing and flehmen of their urine and genitals. If he detects signs that a cow will soon be coming into breeding condition he tries to stop her from leaving the area. His initial advances are aggressively rebuffed, and he may have to herd her for as long as two weeks before she is ready to accept being mated. When it does occur mating is vigorous and prolonged, lasting for half an hour.

PRIVILEGES OF RANK

A male's ability to secure a territory, hold high social rank or beat off rivals is nearly always necessary to persuade females that he is a suitable mate. It is not usually sufficient, however. In territorial antelope and some other species, access to females as mates is completely dependent on the male's having secured a territory. A grey rhebok ram has first to secure a territory, then he can assemble a harem, and only then can be become sexually active. Springbok rams first rise to territorial status at between three and three and a half years old, and then have tenures of 6–20 months during which they have chances to mate. Mountain zebra stallions cannot breed until they have built or taken over a harem, and white rhino bulls who are physically capable of mating when they are five years old can start their reproductive lives only about seven years later, when they secure a territory.

For a lion, hippo, Cape fur seal or rock dassie harem master, or a dominant male baboon, high social rank is necessary to be able to mate, and it is also sufficient. By monopolising access to the females he not only ensures that they mate with no other male, but also that, sooner or later, they will mate with him. With mating assured, courtship is perfunctory, and the long periods of female sexual attractiveness that provoke competition in males of other species may be absent. Among baboons and lions there is a reversal of the usual courtship roles, with the females

Apart from whales and dolphins, hippos are the only African mammals that mate in the water. Mating takes place with the female completely submerged, coming up occasionally for air.

flaunting their sexual receptivity and actively soliciting matings. Elephant cows make the most of their once-in-four-year sexually receptive periods by actively seeking out a high-ranking musth bull to mate with.

Female baboons on heat solicit copulations by presenting their swollen red behinds to likely males, and flashing their white eyelids at them. Although there is a definite dominance hierarchy among a troop's males, a female mates apparently indiscriminately with lower-ranking as well as high-ranking males. Nonetheless, the matings that coincide with her ovulation, which are the most likely to result in conception, are monopolised by the top-ranking males – the lower ranks get sex, but they do not father many offspring. The exceptions that prove the rule are provided by pairs in which close social ties lead to the male being the female's preferred sexual partner. If a female's favoured consort is not one of the top-ranking males the pair will sneak away from the troop for mating.

Cape fur seals provide the most clear-cut example of a harem mating system among southern African mammals. They haul out on land only to reproduce, and during the brief six week mating and pupping season that is all they do. The reproductive territorial bulls guard their territories and harems continuously while living off their stores of accumulated blubber. The bulls haul out in mid-October and fight savagely among themselves to establish small territories on the shore. Females prefer to be close to the sea and so territories low on the shore provide the best supply of females, and these areas are held by the largest, fiercest males. When the females haul out, starting in November, the bulls herd them into harems, and by late November each territorial bull controls a group of females in the final stages of pregnancy. To keep them inside his territory the harem bull aggressively herds straying cows, dragging them bodily back onto his ground – or if that fails, picking them up and throwing them. Five to seven days after she gives birth a female is mated by her harem bull, who then promptly loses interest in her and allows her to return to the sea.

Dominant male baboons are interested only in females at their most fertile. At other times low-ranking males and youngsters are allowed to mate because they will father no offspring.

ELEPHANTS

An elephant female bears a calf only once every four or five years, and comes back into breeding condition only after weaning, or losing, a calf. By the time that she comes on heat, for three to six days, she has been attractive for two and a half weeks, long enough for her loud subsonic rumbles and the smell of her urine to have advertised her condition to all the bulls in an area of hundreds of square kilometres. The mating behaviour of elephant bulls is tied to the phenomenon of musth – a periodic surge in testosterone levels in mature bulls that are in prime physical condition. Although African elephant bulls are able and willing to mate at any stage of the musth cycle, those that are not in musth are usually kept away from any receptive

females by bulls that are. In addition, elephant cows have a strong preference for their mates to be in musth; it is a reliable indicator of mate quality because only bulls in their physical prime can afford its enormous costs in energy and risks of injury. Sexual overtures by a bull that is not in musth are greeted by shrieks of protest by the cow, which serve to attract the attention of any nearby musth bull who will drive the other away.

For an animal as big and heavy as an elephant, mating presents special challenges, especially since a big bull can be three times the weight of a young cow who is mating for the first time. To add to their difficulties the cow's vulva opens downwards between her hind legs, and the mounted bull has to partly crouch with his hind legs bent to get the tip of his penis down low enough for intromission. Once the probings of his highly mobile penis have located the opening of the cow's vagina actual copulation only takes about 45 seconds. Mating may be repeated three or four times over the next 24 hours, but then it will be four or five years before the cow mates again.

Mounting is often part of low-key dominance skirmishes between young male elephants; it is probably this behaviour that gave rise to the myth that elephants mate only in water.

This nearly celibate lifestyle makes observations of mating elephants rather rare, which led to some quaint myths about their mating habits. One was that elephants are so modest that they hide themselves away in dense cover while copulating, and the other was that elephants mate only in water. In fact, elephants are particularly vocal during mating. Excitement spreads though the whole of a receptive female's social group as they keep up a cacophony of rumbling, trumpeting and screaming. Far from being modest, cows on heat will track down and solicit musth bulls. If elephants only mated in water there would be no elephants in arid areas like Damaraland where the only water is in small, shallow water-holes. The myth is given spurious substance by observations of mounting between adolescent bulls as part of bath-time play. Apart from whales, dolphins and dugongs the only southern African mammals that do mate in water are hippos – the female is completely submerged by the bull's weight, and she lifts her head to take an occasional breath.

SPERM COMPETITION

If a female mates with more than one male during her period of receptivity there is competition between the millions of sperm from different males to be the ones that fertilise the ova. If the matings come in quick succession, the male that produces the largest volume of semen has the best chance of being the one whose sperm reaches and fuses with the ova. A southern right whale cow that is in mating condition attracts a retinue of males and she will mate with several of them in quick succession, classic circumstances for sperm competition. In the battle for access to her single ripe ovum the weapon that the males use is sperm volume: southern right whale bulls have the largest testes in the world. Rapid serial matings also occur in territorial antelope like blue wildebeest and impala as females move from one mating territory to the next, but the males cannot adopt the whales' strategy of competition by sperm volume because they need to be able to mate with several females in rapid succession.

For males a prolonged copulation that goes on until his sperm have had a chance to fertilise the ova is another effective means of ensuring that no other male inseminates the female. This is probably one of the reasons why males of species as diverse as rhinos, striped polecats, white-naped weasels and thick-tailed bushbabies copulate for more than 30 minutes at a time. In canids the mating pair stay locked together by the swelling of the male's penis inside the female's vagina to form the copulatory tie, which blocks other matings until the ova have been fertilised. In some species the tie can last for as long as an hour but in African canids, which are under a continual threat of attack from lions, the tie lasts only a few minutes.

In several rodent species, where the threat from predators precludes a protracted coupling, the male's semen coagulates inside the female's vagina, blocking the passage for later arrivals.

The distractions of courtship and copulation can increase the danger from predators, and sexual behaviour is adapted accordingly. In general, the greater the vulnerability, the quicker the mating. The fleeting contact of a springbok coupling, with the female perhaps not even standing still as the male rears up behind her, contrasts sharply with the 30 minute copulations of white rhinos, which are protected by their bulk, and the 90 minute matings of striped polecats and white-naped weasels, which are protected by their stinking anal-gland secretions.

For a male porcupine or hedgehog, mating with a spiny female is a risky business. There is truth in the cliché that they do it carefully, but in addition, the males of both species have an exceptionally long penis in relation to body size. Pangolins have somewhat different difficulties; the female's long thick tail gets in the way of mating in the usual mammal position and so the male mounts the female from the side.

Scent marks of sprayed urine help male and female leopards find each other when the time is ripe for mating.

CATS

For leopards and other cats, and other species whose members live solitarily and far apart, there is a danger that a female might not be able to attract a male for mating at the critical time for conception. In these species, ovulation is induced by mating. To ensure that the female is stimulated strongly enough to trigger the hormonal changes that result in ovulation, the male has small, backward-pointing spines on his penis which increase the intensity of the stimulus to the female's genitals, and account for her aggressive reaction to the male's withdrawal. Cats of all species mate repeatedly to further enhance the stimulation of the female. The repeated matings that induce ovulation in solitary cats were the foundation for lions' apparently profligate sexuality, which now has additional functions that are tied to lions' social

structures. Lions are famous, or notorious, for copulating repeatedly, at a rate of two to four times an hour, over a period of two or three days. This seeming sexual overkill is just one facet of lions' social organisation (*see* Chapter 5), and it plays two main roles. It reduces rivalry between pride males for the immediate gratification of mating, by providing more mating opportunities than any one male can cope with, and it confuses the issue of paternity of the lioness's cubs. The uncertainty over paternity makes the males even-handed in their tolerance and protection of the cubs. If all of the pride males get their share of mating, none of them can behave as if the cubs are not his offspring, which might include killing them to bring their mother back into breeding condition, as happens when new males take over a pride. The lack of rivalry between the pride males benefits the females by maintaining the males' teamwork in defending the cubs against other males bent on infanticide or a pride take-over.

Top, left: Pride males share access to their females, and each treats all the pride's cubs as if they were his own.

Top, right: A female cat's violent reaction to her mate's withdrawal is stimulated by the tiny, backward-pointing spines on his penis.

Above: Lionesses signal their readiness to mate through chemicals in their urine, which males detect by smell and flehmen.

SEASONS OF NEW LIFE

Advanced pregnancy and lactation make such heavy demands on females that in many species breeding is timed so that young are born when food is readily available. For most herbivores this means that births take place early in the rainy season, when the grass is beginning to grow. For predators it means births at about the same time, so that they can take advantage of the increased numbers of young, vulnerable herbivores. If births are to be seasonal, then mating must be seasonal also.

The most conspicuous mating season among terrestrial mammals in southern Africa is that of impalas. In late summer the shortening hours of daylight trigger an increase in the levels of testosterone in adult males, which have been living in

Below, top to bottom: Rutting impala rams horn and scent-mark bushes, build middens of urine and droppings, and roar to proclaim their status.

bachelor groups through the summer. As low-key sparring and horn fencing intensify towards serious conflict, the increase in aggression between the males fragments the bachelor groups. Rams that are in prime physical condition (usually between four and a half and eight and a half years old) begin to stake out small territories of 5-8 hectares. They thrash bushes with their horns, and mark their territories with middens of dung and urine and by rubbing their foreheads on grass and bushes. To advertise their status and occupation of a territory they give loud roars that sound like a mixture of a cough and a belch, and it is this roaring that makes the impala breeding season so noticeable.

Intruding rams are challenged by the resident's snorting, sticking out his tongue, roaring with his neck stretched forward and head tipped back, parading with his penis extended, and threatening with his horns. If an intruder does not withdraw he will be attacked. Battle tactics include charges, clashing horns, and wrestling with horns locked. Fights are short but serious, and injuries and deaths are not uncommon. A loser that cannot flee fast enough will be gored by the winner.

Meanwhile, the herds of females with their young continue to move around in search of food and water, and their movements take them into the rams' territories – having females available for mating is the reason why the rams establish and hold territories in the first place. Once he has a group of females in his area a ram will try to keep them there for as long as possible. When they try to leave he herds them away from the territory boundary by running at them, snorting, roaring and threatening them with his horns. He checks their reproductive condition by licking, sniffing and flehmen of their urine and genitals (*see* Chapter 6), and courts and mates with any that are on heat, mounting repeatedly for about 10 seconds at a time until mating is successful. Once a female has been mated her presence in the territory is of no further advantage to the ram, and he allows her to leave.

Roaring, displaying to other males, making middens, herding females, courting and mating continue around the clock, and territorial rams have no time even to feed. Continuous strenuous activity takes its toll on their body condition, and they become less and less able to drive off challengers who are trying to take over the territory. At the height of the mating season territories change owners on average every eight days.

Towards the drier, western parts of southern Africa, seasonal rainfall and the flush of new growth that it brings become less and less reliable, and the timing of breeding depends less on the time of year and more on the food supply. Unless extremely poor conditions force them out of the area in search of food, territorial springbok rams stay on their territories throughout the year, waiting for the rain-flushed growth of new grass that will bring the ewes into breeding condition. When the ewes are on heat the rams herd them, check their reproductive conditions, court and mate with them.

Besides seasonal changes in food supply, there is an additional selective pressure on prey animals to drop their young within a restricted period – the threat of predation. Being smaller, weaker, slower and less experienced than adults makes

youngsters much more vulnerable to predators, all of which take full advantage of that vulnerability. But predators can eat only so many prey in a certain time, and since hunger is the main motivation for hunting they do not usually kill more than they can eat. (Cases of surplus killing in which many more prey are killed than can be eaten are very rare in nature.) A sudden surge in births of vulnerable babies satiates the local predators; the supply is simply greater than can be consumed in the short time that it is available. If the same number of young were born over a longer period a higher proportion of them would be killed and eaten. In a manifestation of the dilution effect of living in a selfish herd (*see* Chapter 4), simply being one of many reduces an individual youngster's chances of being targeted. Females that drop their young during the peak birth period have a good chance of their young surviving; those that give birth outside the peak period have a very high chance that their offspring will be killed.

The widely held belief that impala ewes delay giving birth until the rains start is mistaken – if the rains are late the first of the lambs to be conceived are resorbed, aborted, or abandoned and eaten by predators. In a dry spring the first lambs that are seen with their mothers are those that were conceived later in the breeding season.

The most closely synchronised births are those of blue wildebeest: 80–90 per cent of the calves are born within three weeks. For births to be so focused in time, there has to be a correspondingly short mating season: once courtship and mating

Top: Fights between male impalas over breeding territories are short but savage.

Above: An impala ram checks a female's reproductive status.

Above and top, right: Blue wildebeest are one of the few species in which the females do not seek secluded cover in which to give birth. The calf is on its feet within a minute and within five minutes it can run and will have followed its mother deep into the safety of the herd.

begin, when the first females come on heat, the territorial bulls' calling, courtship and aggressive displays, and continual testing of cows for receptivity, actually accelerate oestrus in the rest of the population so that nearly all the cows come into heat at about the same time. Both cows and bulls mate with a series of partners, or a pair will mate repeatedly. The bull courts a cow by following her persistently, rearing onto his hind legs and attempting to mount; he is so persistent that a cow who is not yet ready to mate has to run away and lie down to avoid his attentions. When a bull finds a cow who is ready for mating they copulate repeatedly at intervals of less than a minute, a feat of sexual stamina besides which a lion pair's four times an hour seems positively puritan.

In contrast to blue wildebeest, a kudu bull will mate only when his sniffing, licking and flehmen of a cow's urine and genitals tells him that the time is right for conception. Before that he will not mount even if the cow stands in readiness.

KEEPING IT OUT OF THE FAMILY

When social groups have stable memberships and last long enough for youngsters that were born in the group to grow to sexual maturity, there is a need for ways of avoiding inbreeding. To avoid mating with their fathers, fillies of both mountain and Burchell's zebra leave their natal groups before they first breed. Yellow mongoose females mate with males from neighbouring groups. Brown hyaena females mate with itinerant males rather than those within their clan, to whom they are closely related.

In nearly all species, males disperse away from their birthplaces and set up a home range, or join a social group in which it is unlikely that they will encounter females to whom they are closely related. Young male baboons disperse at puberty when they are six to seven years old. Male spotted hyaenas leave their clan when they become independent of their mothers at two to three years old, but their integration into a new clan may take a considerable time, and while it is going on they use their natal clan as a home base. Perhaps the most structured of dispersal patterns is found in South African lesser bushbabies – young males move only east or west from where they were born.

Wild dogs are unusual in that both males and females disperse: small groups of either males or females split off from a pack and roam for hundreds of kilometres searching for an unoccupied hunting range and dispersers of the opposite sex. A new pack is formed when dispersal groups of opposite sexes meet up. The bonds between the alpha male and female are established within a few days at most, and their dispersal companions take on the roles of the fledgling pack's first generation of helpers.

Until 1973 cheetahs bred so poorly in captivity that the species was placed in the IUCN Red Data Book Endangered category, and there was a real fear that captive breeding would never contribute anything to the conservation of cheetahs in the wild. The immediate cause of the poor breeding in captivity was that the female cheetahs did not come into heat. In 1974 the De Wildt Cheetah Research Centre in South Africa discovered why: female cheetahs go through normal reproductive cycles only if they are kept separate from males. Once the females have come into oestrus males can be introduced, and ovulation and conception follow mating as in nature.

The importance to conservation of an understanding of behaviour has no better illustration than the success of the De Wildt Centre. Since cheetah breeding programmes around the world adopted the De Wildt management system their successes in breeding have been so dramatic that there is now an over-supply of captive cheetahs, and in 1986 the South African Red Data Book classification for the species was upgraded to Out of Danger. Now the conservation challenge is to conserve cheetah habitat, and the challenge for ethological research is to find reliable ways of reintroducing them to the wild.

Cheetahs were on the endangered list until an understanding of their mating behaviour enabled them to be bred successfully in captivity.

CONCLUSION

Breeding cheetahs in captivity is only one of the challenges that an understanding of animal behaviour can help us face. Translocating animals without disrupting social structures, targeting pest control only at the individuals that cause problems, and making wild animals safely accessible to the game viewing public are just a few of the others.

Although today's field workers can call on technology that ethology's founders could not even have dreamed of, one thing remains unchanged; a constraint on progress set not by the needs of research, the abilities of researchers or the limits of technology, but by a lack of resources. It is time that the ecotourism industry follows the lead of visionaries like Zimbabwe's Save Valley Conservancy, and begins to invest significantly in research into animals in the wild.

INDEX

Page numbers in **bold** refer to main entries, those in *italics* refer to photographs.